3D PRINTING

3D PRINTING

BY

ELLA WATTS
GRACE HUGHES

KNOWLEDGE BAKERS

ISBN: 978-93-9001-370-8

Published by : **Knowledge Bakers**
Pune, India
Email: info@knowledgebakers.com

PREFACE

The landscape of manufacturing and design is undergoing a profound transformation with the advent of 3D printing technology. From its humble beginnings as a prototyping tool to its current status as a mainstream manufacturing technique, 3D printing has revolutionized how we conceive, create, and interact with physical objects. As we navigate this dynamic and rapidly evolving field, it becomes increasingly crucial to explore the myriad applications, challenges, and opportunities presented by 3D printing.

This book, simply titled "3D Printing," emerges as a comprehensive guide to understanding the intricacies of this transformative technology. Through collaboration with leading experts and practitioners in the field, we endeavor to provide readers with a holistic view of 3D printing – from its fundamental principles to its cutting-edge applications across various industries and disciplines.

As editors of this volume, we have curated a collection of chapters that delve into the theoretical foundations, practical methodologies, and innovative uses of 3D printing. From additive manufacturing techniques to material science advancements, from architectural applications to medical breakthroughs, each chapter offers valuable insights into the diverse facets of 3D printing and its potential to reshape our world.

Furthermore, this book aims to bridge the gap between theory and practice, offering readers actionable guidance on harnessing the full potential of 3D printing technology. Through case studies, tutorials, and real-world examples, we strive to empower researchers, engineers, designers, and enthusiasts to explore new frontiers and push the boundaries of what is possible with 3D printing.

As we embark on this intellectual journey together, we extend our gratitude to the contributors whose expertise and dedication have made this endeavor possible. We hope that this book serves as an indispensable resource for anyone intrigued by the transformative power of 3D printing and inspires further innovation and collaboration in this exciting field.

Ella Watts
Grace Hughes

CONTENTS

PART-2: 3D PRINTING AND MEDICINE

Part-1

3D PRINTING AND ENGINEERING

CHAPTER-1

STEREO LITHOGRAPHY

1. INTRODUCTION

As the oldest additive manufacturing (AM) technology, stereolithography (SLA) was first developed by Dr. Hideo Kodama in 1981. He saw it as a fast and low-cost method of reconstructing models in 3D space as an alternative to holographic techniques [1]. The first commercially available SLA printer was patented in 1986 by Charles W. Hull, who founded 3D Systems Inc. Their aim was to facilitate rapid prototyping of plastic parts [2]. With the development of a variety of processes, SLA has far surpassed its initial applications in modeling and prototyping and can be utilized to manufacture highly complex and individually designed geometries. The material is also no longer limited to conventional polymers, but the fabrication of composites [3] and even metallic [4] or ceramic [5] specimens is possible. However, to date, SLA is only being utilized to structure one material at a time, and when comparing to other additive manufacturing (AM) technologies, typical SLA processes exhibit superior resolution and better surface qualities (see Table 1) but at slower printing times and higher cost.

Table 1. Comparison of average lateral resolution and surface roughness of AM technologies for polymers and of soft lithography.

	Resolution (μm)	Surface roughness (μm)	Sources
SLA	10–150	0.38–0.61	[6, 7, 8]
Material jetting	25–100	0.47–8.44	[6, 9]
Material extrusion	100–400	3.24–42.97	[6, 9]

	Resolution (µm)	Surface roughness (µm)	Sources
Powder bed fusion	50–100	17–105	[7, 10]
Soft lithography	<0.01	<0.001	[9, 11]

While soft lithography is not an AM technology, it merits mentioning as a competing method due to its superior resolution to most SLA systems. It utilizes elastomeric stamps or molds to fabricate 3D structures and consequently does not have the characteristic advantages of 3D printing such as the ability of direct and rapid fabrication from a computer-aided design (CAD) model [12]. Nevertheless, it is often employed as an alternative to SLA in for example microfluidic applications [6].

2. THE STEREOLITHOGRAPHY SETUP

SLA is a vat polymerization method [13], where layers of the liquid precursor in a vat are sequentially exposed to ultraviolet (UV) light and thereby selectively solidified. A photoinitiator (PI) molecule in the resin responds to incoming light and upon irradiation, locally activates the chemical polymerization reaction, which leads to curing only in the exposed areas. After developing the first layer in that manner, a fresh resin film is applied, irradiated, and cured. Thus, the part incrementally grows layer-after-layer [1]. This principle spans all SLA processes, which can be classified according to the direction of incident light or irradiation method, see Figure 1.

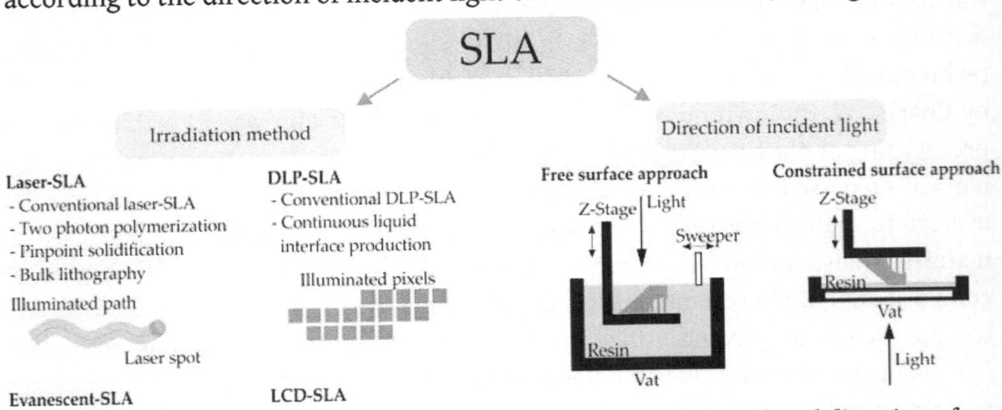

Fig 1. Classification of SLA according to irradiation method (left) and direction of incident light (right).

The required light for solidification of the resin can be applied in two distinct manners; either from above in the free surface approach, or from below through a transparent vat in the constrained surface approach. Irradiation can either be implemented by scanning of each point of the desired cross-section with a laser in laser-SLA or by projecting the entire pixelated image onto the layer in digital light processing (DLP) SLA. A more uncommon method is illumination through an liquid crystal display (LCD) photomask.

These systems and some of their adaptations will be explained in detail in the following sections.

2.1. Free and constrained surface approach

In the **free surface approach**, the building platform on which the printed part grows is situated in a tank of resin and coated with a liquid resin film. Illumination of the desired cross-section, which happens from above the resin bath, cures the first layer. After each layer, the platform with the growing part (i.e. the z-stage), is lowered further into the tank, and new resin is coated on top with a mechanical sweeper. This sets the stage for the subsequent layer [14].

The constrained surface approach or bottom-exposure approach has a building platform, which can be suspended above the resin bath. Illumination from below, through the transparent floor, cures a layer of resin between building platform and vat floor. This layer adheres to the platform, as the z-stage is raised by a defined distance. As each layer is cured, the building platform with the adhered part is elevated, and the part grows suspended from the platform downward [6]. As with the free surface setup, support structures made from the same printing material are needed in case of overhangs and undercuts to ensure adequate adhesion to the platform.

Recently, there has been a trend toward the bottom-exposure approach, as it has certain advantages [6, 15, 16]. The smooth surface, which is created with a narrowly defined layer height due to the precise movement of the z-stage at an accuracy of down to 0.1–1 μm [17, 18], is the main benefit in the constrained surface setup. Without the need for a mechanical sweeper, this layer can be created faster than in the free surface approach, reducing the printing time [15]. Another advantage, which decreases cost, is the lower amount of resin that is needed because the specimen does not have to be completely submerged in the vat [14].

A main disadvantage of the bottom-exposure setup, however, is that attractive forces between printed part and vat floor need to be overcome for each layer [19]. When pulling up the z-stage, the newly cured layer needs to adhere to the layers above, and may not stick to the vat surface. Attempts to reduce this unwanted interaction include the application of hydrophobic coatings of the material tray [15, 18, 19, 20] and modification of the mechanical separation mechanism with tilting steps [21] or application of shear forces [20].

2.2. Laser-stereolithography

In Laser-SLA, also known as vector-based SLA, or often simply referred to as SLA, each layer is cured by scanning of a UV laser onto the resin film. This x-y motion of the laser is implemented by two galvanometers in combination with a dedicated optical system. An example of setup is given schematically in Figure 2. These conventional SLA devices, although more expensive than other AM technologies such as their extrusion-based counterparts, can reach resolutions of 5–10 µm [8].

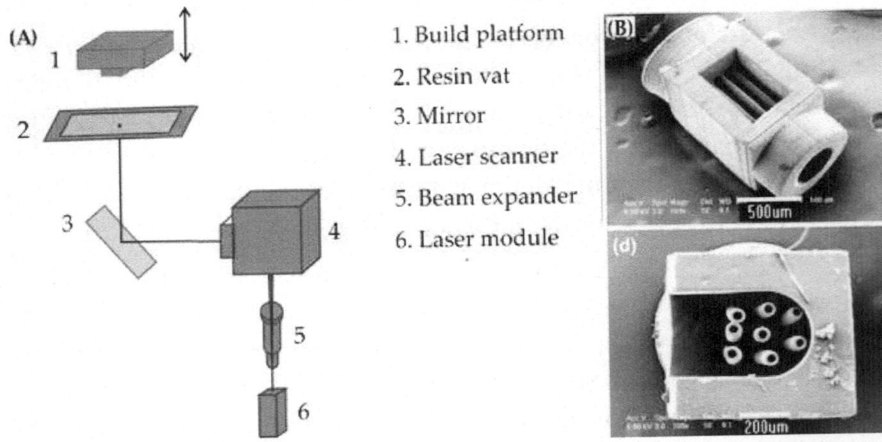

Figure 2. Schematic setup of a laser-SLA printer (A), adapted, and modified from [22]. Scanning electron microscope (SEM) images of a bioreactor with capillaries fabricated with laser-SLA (B), reprinted from [23] by permission of springer nature.

In order to attain these resolutions, a number of parameters need to be considered. Besides accuracy of the z-stage and optimized resin composition (see Section 3.3), which are important factors especially in z-resolution, the manner in which the UV laser is scanned across the desired layer is decisive. The geometry of the precisely defined laser lines, which illuminate the entire cross-section [24] and their accuracy, given by movement of the galvano-mirror, determine lateral resolution. Furthermore, the scanning speed and diameter of the laser spot need to be considered. They, respectively, influence the cure depth and width of the exposed laser lines and thereby affect vertical and lateral resolution [25]. Methods to further improve the resolution to sub-micron regions include two-photon polymerization (TPP) and pinpoint solidification.

2.2.1. Two-photon polymerization (TPP)

TPP was first proposed as an AM method by Strickler et al. [26]. As resolutions superior to 100 nm with surface roughness below 10 nm are attainable [27], it has been extensively studied since then [28], and despite its high-cost, TPP was even commercialized by Nanoscribe GmbH in 2007 [29].

In TPP, excitation of the PI in the resin, and thereby activation of the curing reaction, does not occur in the entire illumination path of the laser, as in conventional SLA, but only in the region of its focal point, called a volume pixel or voxel [30]. A high-intensity femtosecond pulsed laser can cause molecules to absorb two photons simultaneously. As the probability of this phenomenon is proportional to the squared intensity of the laser pulse, the process is limited to the focal point of the laser. Instead of UV light, a laser at twice the wavelength (i.e. half the energy) with near-infrared (NIR) light such as a Titanium-sapphire laser is employed in TPP. The energy necessary for excitation is nevertheless attained by the combination of the energies of both individual photons [31].

As portrayed in Figure 3, the spatially constrained 3D voxel in TPP allows for curing of shapes inside the resin bath and not just on its surface. This eliminates the need for layer-wise production and enables fabrication of extremely complex geometries including freely moving parts without superfluous support structures [28].

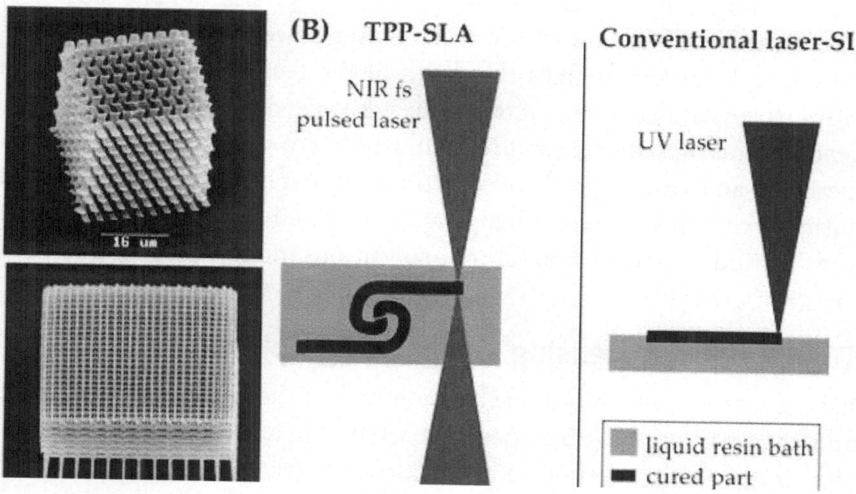

Figure 3. SEM images of microstructures fabricated with TPP (A), reprinted with permission from [32] (Ovsianikov A. et al. ultra-low shrinkage hybrid photosensitive material for two-photon polymerization microfabrication.), copyright (2008) American Chemical Society. Scheme of TPP vs. conventional laser-SLA (B) redrawn and adapted from [33].

One of the challenges which remain in TPP is the restriction to extremely small geometries in the mm range [34] and low writing speed of the laser lines. At a maximum of a few mm/s, it cannot compare to hundreds of mm/s, which is attainable with conventional laser-SLA methods [8]. Developing a suitable PI could help speed up the process. Conventional UV initiators have the drawback of low-activity in TPP. In order to augment their response, the design of molecules with specific structures is necessary [35].

2.2.2. Pinpoint solidification

A method similar to TPP, pinpoint solidification, was proposed by Ikuta et al. in 1998 under the name of super integrated hardened polymer SLA (Super IH) process [36]. A tightly focused laser is used and as with TPP, due to the high intensity in the focal point of the laser, curing of the resin can only be achieved in this voxel. The mechanism, however, is that of conventional single photon polymerization. Thus, resolutions of below 0.4 μm have been reached without the use of expensive fs pulsed lasers [37]. This process has as of yet not been commercialized, and very little research is invested in pinpoint solidification-SLA.

2.2.3. Bulk lithography

In bulk lithography, 3D textures can be created by variation of exposure energy. The cure depth, which is a direct function of laser power or scan velocity (i.e. of the applied energy), thereby defines the depth of the features [38]. One can thus see the entire part as only existing of one layer with varying thickness. This eliminates the sometimes abrupt steps in z-direction, which are generated with conventional SLA methods, and vastly speeds up printing. Although this process is not capable of printing structures with overhangs and is limited to geometries thinner than 0.25 mm, it could have potential future applications in high-throughput fabrication of microstructures [39].

2.3. Digital light processing stereolithography

DLP is a method, which can reach resolutions in the order of 25 μm [7]. Smallest feature sizes of 0.6 μm have also been reported [40], and resins filled with ceramic particles have been printed via DLP with layer heights of 15 μm and with lateral resolutions of 40 μm [18].

In contrast to laser-SLA, the entire cross-section of a layer is illuminated simultaneously by a DLP light engine, as shown in Figure 4. The digital micromirror device (DMD) is the key component and functions as a dynamic mask for the DLP process. It is constructed of an array of mirrors, each one representing a single pixel. Individual tilting of every mirror enables fast and reliable switching of pixels [42]. When linked with a computer for image processing, a light source (often LED), and optics, it can project desired cross-sections of light quickly and precisely [42]. The fast switching speed of the DMD is a prerequisite for realizing grayscale illumination which can be beneficial for precise control over exposure time and by extent energy dosage [40].

With its pixel-based exposure mechanism, DLP is excellent for illumination of sharp corners but can cause saw-tooth type surface roughness on otherwise curved surfaces [18]. Consequently, when aiming for higher resolution, the pixel size needs to be reduced with the help of designated optics. As the DMD has a fixed amount of mirrors, this leads to shrinkage of the image and reduces maximum geometry size. Large parts are thus often printed at lower resolutions than small ones. While not quite reaching the sub-micron resolutions of laser-SLA, DLP retains the advantages of lower cost and higher printing speeds [40, 43].

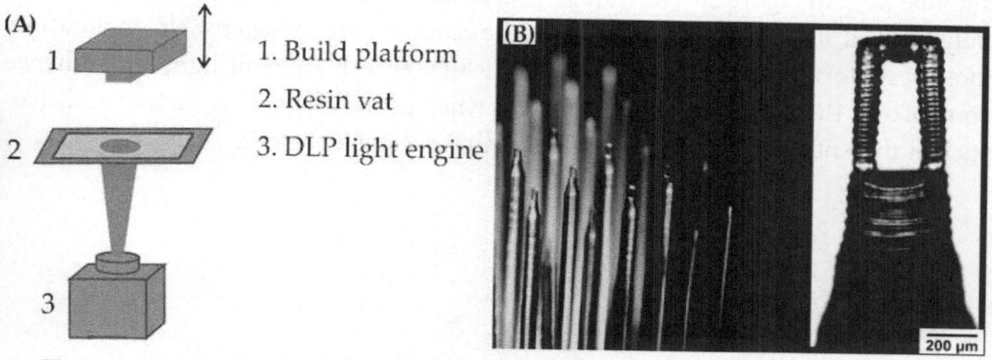

Figure 4. Setup of a DLP-SLA printer (A) adapted and redrawn from [22]. Printed polymer structures (B) adapted and reprinted with permission from [41] (Macdonald NP. et al., 3D printed micrometer-scale polymer mounts for single crystal analysis). Copyright (2017) American Chemical Society.

2.3.1. Continuous liquid interface production (CLIP)

Continuous liquid interface production (CLIP) is a type of constrained surface DLP process, where a thin film between building platform and the material tray is not cured and remains liquid. The so-called dead zone at the interface can be generated by utilizing a vat with a floor that is permeable to oxygen. This inhibits curing, and the resin in contact with oxygen remains liquid. Recoating mechanisms are thereby superfluous and continuous elevation of the building platform can be achieved, which improves surface quality and drastically increases printing speed up to 500 mm/h [44]. Similarly, to attain a liquid interface film, a high-density inert and an immiscible liquid layer such as brine has been proposed [45]. CLIP has been commercialized by Carbon Inc., and is establishing itself in the AM market due to its reduced printing times [46].

2.4. Liquid crystal display stereolithography

Since its development in 1997 by Bertsch *et al.* [47], using an LCD device as a dynamic mask for SLA has been almost completely replaced by the DLP counterpart. The latter benefits are from superior switching speeds at higher accuracy [48]. Nevertheless, it merits mentioning as a low-cost alternative to DLP with commercially available LCD printers primarily catering to the laypersons demographic as opposed to the industry [49].

3. RESINS IN SLA

Photocurable resins for SLA all have the same essential components, as summarized in Figure 5. The liquid precursors, which form the network when polymerized, as well as PIs, which start the reaction, are indispensable. In addition, most resin formulations have inert dyes, which absorb incident light and enhance control over the polymerization. Especially when using filled resins, further additives such as diluents, surfactants, or other stabilizers can be present.

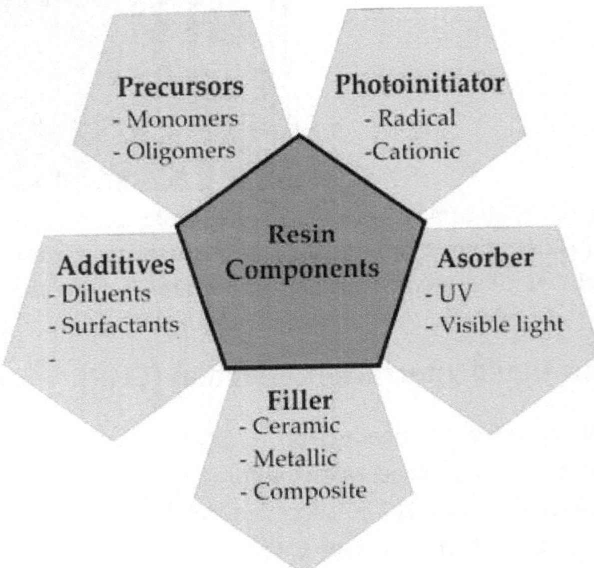

Figure 5. Resin components in SLA.

3.1. Precursors

The precursors in SLA are liquid molecules, which can be linked together (i.e. polymerized), after exposure to light to form a solid 3D network. Depending on the future application and desired attributes, a variety of monomers, oligomers, or prepolymers can be utilized.

Acrylate-based resins are common in the SLA process, as they exhibit high reactivities, which is advantageous for fast building speeds [7]. Different types of acrylates are readily available to tune mechanical properties and thermal resistance for example by altering the number of reactive groups [8] or by employing different oligomers such as urethane acrylates [50]. One disadvantage of acrylate resins is their high-shrinkage during printing, causing potential distortion of the printed part.

As a solution, the combination with methacrylates is often implemented [51]. The resin's sensitivity to oxygen, which inhibits the polymerization reaction, is another challenge.

Epoxy systems have a different curing mechanism than acrylates. They are based on cationic rather than radical photopolymerization and need longer reaction times, are inhibited by moisture, but have the advantage of stability against oxygen [52]. Additionally, epoxy resins exhibit significantly lower shrinkage than their acrylate counterparts [53]. In order to exploit the advantages of both alternatives, hybrid systems have been created. Combination of acrylate and epoxy-based resins lead to fast curing, low-shrinkage materials and are nowadays the standard in most commercial systems [54, 55].

3.2. Photoinitiators (PIs)

The PI is the resin component, which reacts to light. Once irradiated at the correct wavelength, it is excited and can initiate the curing reaction. A suitable PI, depending on the nature of utilized precursor needs to be selected. Type and amount of PI can substantially influence reaction kinetics, necessary light dosage, conversion, cross-linking density, and by extent, mechanical properties of the printed parts [8, 56, 57].

3.3. Absorbers

Another component that is essential in most SLA processes is a light absorber, which reduces the penetration of light into the resin and limits the depth until, which the resin is cured. Especially for complex geometries with undercuts, this cure depth needs to be precisely defined in order to prevent excessive curing in z-direction and loss in feature development [14, 17, 40]. The most commonly used UV absorbers are benzotriazole derivatives [58].

3.4. Filled resins

Fabrication of metal or ceramic materials via SLA has been implemented by filling resins with powder, printing the parts, and subsequently debinding and sintering the printed specimens [4, 5], as shown in Figure 6. During debinding, the organic resin components are removed by pyrolysis. This binder burnout is easier for thin structures with high filler content, as otherwise, defects such as cracks can form [5]. In the subsequent sintering step, the metal or ceramic powder, which remains, is further thermally treated to achieve dense structures [21]. In order to attain geometrically accurate parts, the material specific shrinkage coefficients need to be taken into account, and high filler content is beneficial to reduce shrinkage [60]. Variations of the thermal treatment include implementing an additional drying step prior to debinding to remove solvents or combining debinding and sintering into one single but correspondingly longer thermal process to eliminate potential defect sources during transportation of the fragile brown parts [59].

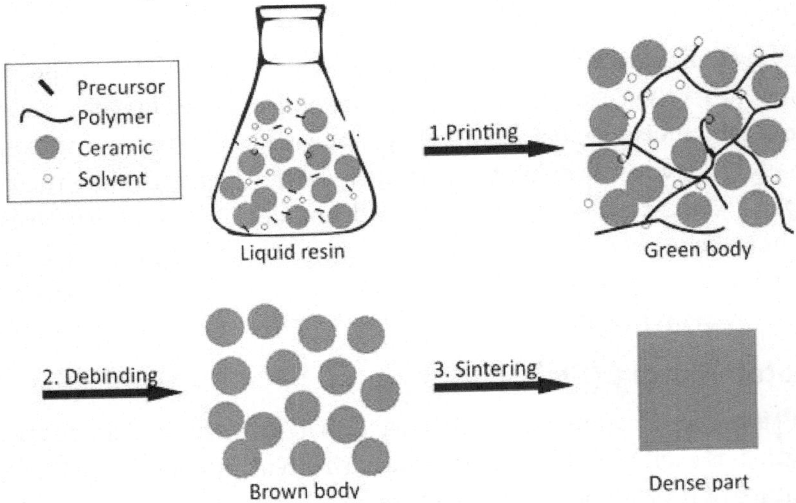

Figure 6. Scheme of the formation of dense ceramic components from filled resins by SLA, redrawn and adapted from [59].

Particles smaller than the layer height need to be utilized, and as previously mentioned, maximum particle content is desirable. In highly filled resins with particles in size range of the wavelength of light, scattering is the main interaction mechanism with light and consequently determines cure depth and affects resolution. Reducing the refractive index difference between filler and matrix is a common approach to minimizing scattering [61]. It is noteworthy that composite materials, where the organic component is not removed but retains its matrix function in the final part, have also been manufactured with SLA [3, 62]. By adding (nano)-particles to the SLA resin, mechanical, thermal, optical, or even electrical properties can be further amended [63, 64, 65].

3.5. Additives

A high-volume fraction of solid loading can cause certain disadvantages. Especially for smaller particles with the large surface area, the viscosity of the slurry rises with particle content [4]. This changes the flow behavior of the resin, interferes with coating mechanisms, and increases the mechanical force necessary for the elevation of the building platform in constrained surface setups [19]. Approaches to reduce viscosity include the application of temperature [66], the use of diluents [67], or evoking shear thinning behavior [68].

Rheological additives and stabilizers can increase solid loading and are necessary for extended shelf-life of slurries as well as for stability during longer printing jobs

[68]. Agglomeration and sedimentation of particles need to be avoided to ensure continuous, homogeneous ceramic or metal powder distribution. To that end, dispersants such as oligomeric surfactants [69], long chained acids like oleic acid [70], or phosphine oxides with aliphatic chains [71] have been used.

3.6. Post-processing

After removal of the built part from the platform, any support structures that had been necessary for the printing process need to be cut from the green part. Cleaning in suitable solvents and drying of the structure is often followed by sanding of support residues. Post-curing in a UV chamber can be implemented to complete conversion of the polymerization reaction and thereby attain improved mechanical properties [72]. In the case of filled resins, debinding and sintering are the final post-processing steps.

4. APPLICATIONS

SLA is a very versatile method with applications in a variety of industries. The aerospace and automotive industries can, for instance, benefit from rapid manufacturing of high-performance materials. Microfluidics and medicine are furthermore, significant fields where SLA shows great potential and is already being applied successfully.

Fully polymeric materials structured by SLA can range in their properties from highly elastic silicones for applications in soft robotics [45] to high-strength thermally post-cured epoxy resins [73]. Their limited thermo-mechanical stability is, however, an issue for most polymeric materials. Using filled resins to create metal or ceramic structures, is a possibility in SLA, as previously mentioned in Section 3.4. Furthermore, polymer-derived ceramics can be manufactured by using monomers as precursors, which contain the essential components to form ceramics upon pyrolysis. These methods offer superior versatility in geometry than casting or machining processes and can yield components for high-temperature applications such as in propulsion systems or as thermal insulators [74].

Recently, SLA has been extensively investigated in the field of microfluidics, where small fluid volumes need to be precisely manipulated through micro-sized channels for applications such as inkjet print heads or lab-on-a-chip technologies [6]. When compared to material extrusion and jetting, DLP-SLA shows superior resolution, smaller possible feature sizes, reduced surface roughness, and faster production times in this application [75]. Channels with dimensions below 100 µm and valves, pumps, as well as multiplexers for mixing [76], can be fabricated rapidly and easily.

Applications in medicine, where patient-specific designs are often necessary to accommodate for individual anatomies, can greatly benefit from AM as well. Some examples are depicted in Figure 7. CT or MRI scans can be employed to determine the geometrical specifications, from which devices are then manufactured. Craniofacial implants out of porously structured hydroxyapatite have for example been implanted in patients with large bone defects [80].

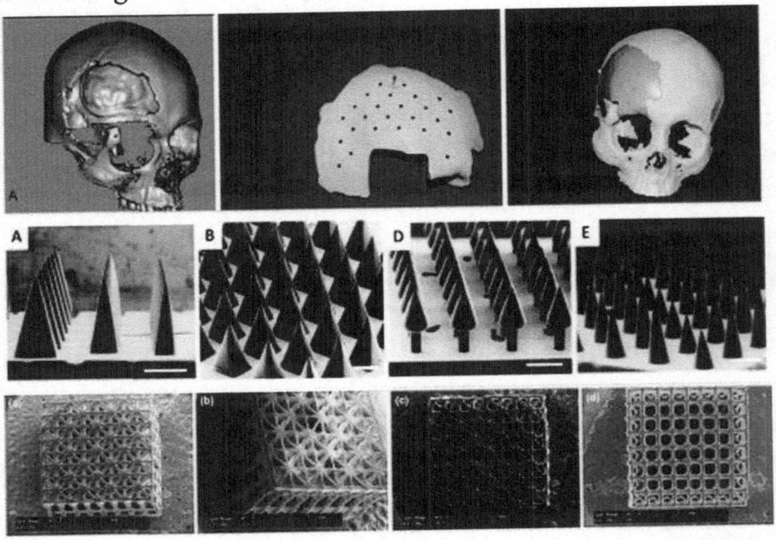

Figure 7. Top row: CAD model of skull defect (A), SLA fabricated cranial implant (C), and an implant placed into skull model (D), reprinted and modified from [77]. Middle row: SEM images of SLA printed microneedle structures for transdermal drug delivery, modified and reprinted from [78]. Bottom row: SEM images of tissue engineering scaffolds for bone regeneration by SLA, modified and reprinted from [79] by permission of Springer Nature.

In dentistry, CAD modeling has been applied since the 70s in the creation of crowns, which are used to cover a damaged tooth, and dentures, which are removable or fixed devices to replace lost teeth [81]. Now, many AM technologies including SLA can be employed to speed up the process between the acquisition of the geometrical data and implantation of the device into the patient [82]. A second application in healthcare, where AM has become the norm is the manufacturing of hearing aids. SLA can reduce the manufacturing time of these custom-made devices from more than a week to less than a day while also improving wear comfort [83, 84].

Medical applications of SLA are not limited to the fabrication of implants, prostheses, or other medical devices, but drug delivery systems such as micro-needles, capable of administering drugs by painlessly penetrating the skin [78], or 3D printed tablets for individual dosage regulation [85] have also been established. SLA can furthermore be used in medical imaging to create 3D models [86] for preoperative planning [87] or educational purposes [88]. 3D cell culture for more accurate in vitro models to study diseased, as well as healthy tissues, can likewise be created [89]. Tissue engineering constructs in regenerative medicine have also been fabricated with SLA methods [90] and even bioprinting, where live cells are incorporated into the printed scaffold is currently being investigated [91].

5. COMMERCIALLY AVAILABLE SLA SYSTEMS

A plethora of SLA printers are currently on the market, ranging in price and quality from amateur desktop applications to professional high-precision machines. A brief overview of their specifications is given in Table 2.

Table 2. Comparison of commercially available SLA systems.

SLA-type	Lateral resolution (µm)	Printing speed (mm/h)	Maximum print size (mm)	Commercial vendors	Sources
Laser	6–140	14	27–750	3DSystems, Formlabs, XYZPrinting,	[92, 93, 94, 95]
TPP	0.400	—	100 × 100 × 3	Nanoscribe	[34]
DLP	33–120	25–150	45–230	EnvisionTEC, Kudo3D	[96, 97, 98]
CLIP	50–100	500	80–320	Carbon	[44]
LCD	50–100	20–60	55–160	SparkMaker, Photocentric	[49, 99]

LCD-SLA is only available as a low-budget 3D printer for hobbyists. The more common methods such as conventional laser-SLA and DLP-SLA, however, have their low-cost and low-resolution desktop editions for consumers but can also offer high-end methods for professionals and industrial applications. While CLIP and TPP are both in higher price ranges, the former is designed for the extremely fast production of larger parts, while the latter is used for very slow fabrication of small parts with sub-micron resolution.

6. OUTLOOK

As the front-runner for high-resolution 3D printing, SLA retains substantial limitations due to the often high costs of this AM method, which are augmented further by slow printing velocities. Additionally, the fact that printing is only possible with one resin at a time severely restricts potential applications.

Trends toward faster manufacturing have already been set by the CLIP technology [44]. Other continuous DLP methods, where building platforms are raised at a constant rate during the printing process, have also been established [76]. For laser-SLA processes, illumination itself is a limiting factor in reducing printing times and different approaches to increase throughput are necessary. Using a broader scanning pattern for bulk features and applying more precise, narrow lines only in areas where the maximum resolution is required, such as for fine structures and at surfaces, is one method, which is already being implemented [93]. The development of hybrid systems of DLP and laser techniques is currently being investigated as well. Similarly, the inner area should be illuminated via pixel-based DLP, and only round surfaces drawn with the vector-based system of laser-SLA. This could further reduce printing times to rates comparable to DLP while maintaining high-accuracy of laser illumination [22].

Another method, which combines laser and DLP-SLA, addresses the compromise between build size and resolution in DLP. A proposition to retain small pixels and thereby high-resolution even when printing large parts is to laterally stitch the projected images. If a layer has a cross-section exceeding the attainable size by the DMD, it can be divided into smaller areas, which are then illuminated one after the other [100]. This combination of scanning and projection-based illumination, also called large area projection micro SLA (LAPμLA) [101], can lead to the low-cost fabrication of cm-sized objects with a μm-range resolution [102].

Modification of available SLA systems to manufacture parts from multiple materials has been attempted. This usually includes a time consuming cleaning step between material changes. Thus, minimum feature sizes and resolution are no more comparable to conventional SLA than required printing time [103, 104, 105]. Nevertheless, after thorough investigation and development, these methods could help to further extend the application spectrum of SLA in the future.

REFERENCES

1. Kodama H. Automatic method for fabricating a three-dimensional plastic model with photo-hardening polymer. The Review of Scientific Instruments. 1981;52(11):1770-1773
2. Hull CW. Apparatus for production of three-dimensional objects by stereolithography. 1984. US4575330B1
3. Wang X, Jiang M, Zhou Z, Gou J, Hui D. 3D printing of polymer matrix composites: A review and prospective. Composites. Part B, Engineering. 2017;110:442-458
4. Bartolo PJ, Gaspar J. Metal filled resin for stereolithography metal part. CIRP Annals. 2008;57(1):235-238

5. Halloran JW. Ceramic Stereolithography: Additive manufacturing for ceramics by photopolymerization. Annual Review of Materials Research. 2016;46(1):19-40

6. Waheed S, Cabot JM, Macdonald NP, Lewis T, Guijt RM, Paull B, et al. 3D printed microfluidic devices: Enablers and barriers. Lab on a Chip. 2016;16(11):1993-2013

7. Ligon SC, Liska R, Stampfl J, Gurr M, Mülhaupt R. Polymers for 3D printing and customized additive manufacturing. Chemical Reviews. 2017;117(15):10212-10290

8. Stampfl J, Baudis S, Heller C, Liska R, Neumeister A, Kling R, et al. Photopolymers with tunable mechanical properties processed by laser-based high-resolution stereolithography. Journal of Micromechanics and Microengineering. 2008;18(12):125014

9. Lee JM, Zhang M, Yeong WY. Characterization and evaluation of 3D printed microfluidic chip for cell processing. Microfluidics and Nanofluidics. 2016;20(1):5

10. Launhardt M, Wörz AW, Loderer A, Laumer T, Drummer D, Hausotte T, et al. Detecting surface roughness on SLS parts with various measuring techniques. Polym Test. 2016;53:217-226

11. Quake SR, Scherer A. From micro- to nanofabrication with soft materials. Science. 2000 Nov 24;290(5496):1536-1540

12. Xia Y, Whitesides GM. Soft lithography. Angewandte Chemie International Edition. 1998;37(5):550-575

13. ASTM ISO/ASTM52921-13 Standard Terminology for Additive Manufacturing-Coordinate Systems and Test Methodologies. West Conshohocken: ASTM International; 2013

14. Han L-H, Mapili G, Chen S, Roy K. Projection microfabrication of three-dimensional scaffolds for tissue engineering. Journal of Manufacturing Science and Engineering. 2008;130(2):21005

15. Pan Y, Zhou C, Chen Y. A fast mask projection Stereolithography process for fabricating digital models in minutes. Journal of Manufacturing Science and Engineering. 2012;134(5):51011

16. Huang Y-M, Kuriyama S, Jiang C-P. Fundamental study and theoretical analysis in a constrained-surface stereolithography system. International Journal of Advanced Manufacturing Technology. 2004;24(5-6):361-369

17. Choi J, Wicker RB, Cho S, Ha C, Lee S. Cure depth control for complex 3D microstructure fabrication in dynamic mask projection microstereolithography. Rapid Prototyping Journal. 2009;15(1):59-70

18. Hatzenbichler M, Geppert M, Seemann R, Stampfl J. Additive manufacturing of photopolymers using the Texas Instruments DLP lightcrafter. In: Proceedings of SPIE, Emerging Digital Micromirror Device Based Systems and Applications V. San Francisco: International Society for Optics and Photonics; 2013. pp. 86180A

19. Huang Y-M, Jiang C-P. On-line force monitoring of platform ascending rapid prototyping system. Journal of Materials Processing Technology. 2005;159(2):257-264

20. Chen Y, Zhou C. Digital mask-image-projection-based additive manufacturing that applies shearing force to detach each added layer. 2013. US13872954

21. Schwentenwein M, Homa J. Additive manufacturing of dense alumina ceramics. International Journal of Applied Ceramic Technology. 2015;12(1):1-7

22. Busetti B, Lutzer B, Stampfl J. Development of a hybrid exposure system for lithographybased additive manufacturing technologies. In: Laser 3D Manufacturing V. San Francisco: International Society for Optics and Photonics; 2018. p. 4

23. Xia C, Fang NX. 3D microfabricated bioreactor with capillaries. Biomedical Microdevices. 2009;11(6):1309-1315

24. Lee IH, Cho DW. Micro-stereolithography photopolymer solidification patterns for various laser beam exposure conditions. International Journal of Advanced Manufacturing Technology. 2003;22(5-6):410-416

25. Zhang X, Jiang XN, Sun C. Micro-stereolithography of polymeric and ceramic microstructures. Sensors and Actuators A: Physical. 1999;77(2):149-156

26. Denk W, Strickler JH, Webb WW. Two-photon laser scanning fluorescence microscopy. Science. 1990;248(4951):73-76

27. Takada K, Sun H-B, Kawata S. Improved spatial resolution and surface roughness in photopolymerization-based laser nanowriting. Applied Physics Letters. 2005;86(7):71122

28. Park S-H, Yang D-Y, Lee K-S. Two-photon stereolithography for realizing ultraprecise three-dimensional nano/microdevices. Laser & Photonics Reviews. 2009;3(1-2):1-11

29. Das Unternehmen im Detail – Nanoscribe GmbH [Internet]. [cited 2018 Mar 10]. Available from: https://www.nanoscribe.de/de/unternehmen/

30. Serbin J, Egbert A, Ostendorf A, Chichkov BN, Houbertz R, Domann G, et al. Femtosecond laser-induced two-photon polymerization of inorganic-organic hybrid materials for applications in photonics. Optics Letters. 2003;28(5):301-303

31. Maruo S, Nakamura O, Kawata S. Three-dimensional microfabrication with two-photon-absorbed photopolymerization. Optics Letters. 1997;22(2):132

32. Ovsianikov A, Viertl J, Chichkov B, Oubaha M, MacCraith B, Sakellari I, et al. Ultra-low shrinkage hybrid photosensitive material for two-photon polymerization microfabrication. ACS Nano. 2008;2(11):2257-2262

33. Wu S, Serbin J, Gu M. Two-photon polymerisation for three-dimensional micro-fabrication. Journal of Photochemistry and Photobiology A: Chemistry. 2006;181(1):1-11

34. Data sheet: Photonic Professional G T [Internet]. [cited 2018 Mar 14]. Available from: https://www.nanoscribe.de/files/2715/1912/9082/DataSheet_Photonic_Professional_GT.pdf

35. Pucher N, Rosspeintner A, Satzinger V, Schmidt V, Gescheidt G, Stampfl J, et al. Structure-activity relationship in D-π-A-π-D-based photoinitiators for the two-photon-induced photopolymerization process. Macromolecules. 2009;42(17):6519-6528

36. Ikuta K, Maruo S, Kojima S. New micro stereo lithography for freely movable 3D micro structure-super IH process with submicron resolution. In: Proceedings of the IEEE International Conference on Micro Electro Mechanical Systems (MEMS). Heidelberg: IEEE; 1998. p. 290-295

37. Maruo S, Ikuta K. Submicron stereolithography for the production of freely movable mechanisms by using single-photon polymerization. Sensors and Actuators A: Physical. 2002;100(1):70-76

38. Gandhi P, Bhole K. Characterization of "bulk lithography" process for fabrication of three-dimensional microstructures. Journal of Micro and Nano-Manufacturing. 2013;1(4):41002

39. Gandhi P, Bhole K, Chaudhari N. Fabrication of textured 3D microstructures using "bulk lithography". In: ASME 2012 International Manufacturing Science and Engineering Conference. ASME; 2012. pp. 959

40. Sun C, Fang N, Wu DM, Zhang X. Projection micro-stereolithography using digital micro-mirror dynamic mask. Sensors and Actuators A: Physical. 2005;121(1):113-120

41. Macdonald NP, Bunton GL, Park AY, Breadmore MC, Kilah NL. 3D printed micrometer-scale polymer mounts for single crystal analysis. Analytical Chemistry. 2017;89(8):4405-4408

42. Katal G, Tyagi N, Joshi A. Digital light processing and its future applications. International Journal of Scientific and Research Publications. 2013;3(1):2250-3153

43. Baumgartner S, Pfaffinger M, Busetti B, Stampfl J. Comparison of Dynamic Mask- And Vector-Based Ceramic Stereolithography. In: Proceedings of the 41st International Conference on Advanced Ceramics and Composites. Hoboken: Wiley; 2018. P. 163-173

44. Tumbleston JR, Shirvanyants D, Ermoshkin N, Januszwiewiez R, Johnsn AR, Kelly D, et al. Continuous liquid interface production of 3D objects. Science. 2015;347(6228):1349-1352

45. Thrasher CJ, Schwartz JJ, Boydston AJ. Modular elastomer Photoresins for digital light processing additive manufacturing. ACS Applied Materials & Interfaces. 2017;9(45):39708-39716

46. Process – Carbon [Internet]. [cited 2018 Mar 11]. Available from: https://www.carbon3d.com/process/

47. Bertsch A, Zissi S, Jézéquel JY, Corbel S, André JC. Microstereophotolithography using a liquid crystal display as dynamic mask-generator. Microsystem Technologies. 1997 Feb 21;3(2):42-47

48. Hornbeck LJ. The DMDTM Projection Display Chip: A MEMS-Based Technology. MRS Bull. 2001;26(4):325-327

49. SparkMaker [Internet]. [cited 2018 Mar 11]. Available from: https://www.sparkmaker3d.com/

50. Coats AL, Harrison JP, Hay JS, Ramos Manuel Jacinto. Stereolithography resins and methods. 2003. US10628304

51. Murphy EJ, Ansel RE, Krajewski JJ. Method of forming a three-dimensional object by stereolithography and composition therefore. 1989. US07429568

52. Sipani V, Scranton AB. Kinetic studies of cationic photopolymerizations of phenyl glycidyl ether: Termination/trapping rate constants for iodonium photoinitiators. Journal of Photochemistry and Photobiology A: Chemistry. 2003;159(2):189-195

53. Esposito Corcione C, Greco A, Maffezzoli A. Photopolymerization kinetics of an epoxy-based resin for stereolithography. Journal of Applied Polymer Science. 2004;92(6):3484-3491

54. Lee TY, Carioscia J, Smith Z, Bowman CN. Thiol-Allyl Ether-Methacrylate Ternary Systems. Evolution Mechanism of Polymerization-Induced Shrinkage Stress and Mechanical Properties. Macromolecules. 2007;40(5):1473-1479

55. Zhiwei G, Jianhua M, Shuhuai H, Hongquan X. Development of a hybrid photopolymer for stereolithography. Journal of Wuhan University of Technology-Materials Science Edition. 2006;21(1):99-101

56. Badev A, Abouliatim Y, Chartier T, Lecamp L, Lebaudy P, Chaput C, et al. Photopolymerization kinetics of a polyether acrylate in the presence of ceramic fillers used in stereolithography. Journal of Photochemistry and Photobiology A: Chemistry. 2011;222(1):117-122

57. Bail R, Patel A, Yang H, Rogers CM, Rose FRAJ, Segal JI, et al. The effect of a type I photoinitiator on cure kinetics and cell toxicity in projection-microstereolithography. Procedia CIRP. 2013;5:222-225

58. Bail R, Hong JY, Chin BD. Effect of a red-shifted benzotriazole UV absorber on curing depth and kinetics in visible light initiated photopolymer resins for 3D printing. Journal of Industrial and Engineering Chemistry. 2016;38:141-145

59. de Blas Romero A, Pfaffinger M, Mitteramskogler G, Schwentenwein M, Jellinek C, Homa J, et al. Lithography-based additive manufacture of ceramic biodevices with design-controlled surface topographies. International Journal of Advanced Manufacturing Technology. 2017;88(5-8):1547-1555

60. Hinczewski C, Corbel S, Chartier T. Ceramic suspensions suitable for stereolithography. Journal of the European Ceramic Society. 1998;18(6):583-590

61. Griffith ML, Halloran JW. Scattering of ultraviolet radiation in turbid suspensions Prediction of ceramic stereolithography resin sensitivity from theory and measurement of diffusive photon Scattering of ultraviolet radiation in turbid suspensions. Journal of Applied Physics. 1997;81(47):2538-24902

62. Manjubala I, Woesz A, Pilz C, Rumpler M, Fratzl-Zelman N, Roschger P, et al. Biomimetic mineral-organic composite scaffolds with controlled internal architecture. Journal of Materials Science: Materials in Medicine. 2005;16(12):1111-1119

63. Farahani RD, Dubé M, Therriault D. Three-dimensional printing of multifunctional nanocomposites: Manufacturing techniques and applications. Advanced Materials. 2016;28(28):5794-5821

64. Lin D, Jin S, Zhang F, Wang C, Wang Y, Zhou C, et al. 3D stereolithography printing of graphene oxide reinforced complex architectures. Nanotechnology. 2015;26(43):434003

65. Credi C, Fiorese A, Tironi M, Bernasconi R, Magagnin L, Levi M, et al. 3D printing of cantilever-type microstructures by stereolithography of ferromagnetic photopolymers. ACS Applied Materials & Interfaces. 2016;8(39):26332-26342

66. Tomeckova V, Halloran JW. Flow behavior of polymerizable ceramic suspensions as function of ceramic volume fraction and temperature. Journal of the European Ceramic Society. 2011;31(14):2535-2542

67. Melchels FPW, Feijen J, Grijpma DW. A review on stereolithography and its applications in biomedical engineering. Biomaterials. 2010;31(24):6121-6130

68. De Hazan Y, Heinecke J, Weber A, Graule T. High solids loading ceramic colloidal dispersions in UV curable media via comb-polyelectrolyte surfactants. Journal of Colloid and Interface Science. 2009;337(1):66-74

69. Teng WD, Edirisinghe MJ, Evans JRG. Optimization of dispersion and viscosity of a ceramic jet printing ink. Journal of the American Ceramic Society. 1997;80(2):486-494

70. Li K, Zhao Z. The effect of the surfactants on the formulation of UV-curable SLA alumina suspension. Ceramics International. 2017;43(6):4761-4767

71. Goswami A, Ankit K, Balashanmugam N, Umarji AM, Madras G. Optimization of rheological properties of photopolymerizable alumina suspensions for ceramic microstereolithography. Ceramics International. 2014;40(2):3655-3665

72. Kim H-C, Lee S-H. Reduction of post-processing for stereolithography systems by fabrication-direction optimization. Computer Design. 2005;37(7):711-725

73. Kuang X, Zhao Z, Chen K, Fang D, Kang G, Qi HJ. High-Speed 3D Printing of High-Performance Thermosetting Polymers via Two-Stage Curing. Macromol Rapid Commun. 2018;39(7):1700809

74. Eckel ZC, Zhou C, Martin JH, Jacobsen AJ, Carter WB, Schaedler TA. Additive manufacturing of polymer-derived ceramics. Science. 2016;351(6268):58-62

75. Macdonald NP, Cabot JM, Smejkal P, Guijt RM, Paull B, Breadmore MC. Comparing microfluidic performance of three-dimensional (3D) printing platforms. Analytical Chemistry. 2017;89(7):3858-3866

76. Gong H, Woolley AT, Nordin GP. High density 3D printed microfluidic valves, pumps, and multiplexers. Lab on a Chip. 2016;16(13):2450-2458

77. Kumta S, Kumta M, Jain L, Purohit S, Ummul R. A novel 3D template for mandible and maxilla reconstruction: Rapid prototyping using stereolithography. Indian Journal of Plastic Surgery. 2015;48(3):263-273

78. Johnson AR, Caudill CL, Tumbleston JR, Bloomquist CJ, Moga KA, Ermoshkin A, et al. Single-step fabrication of computationally designed microneedles by continuous liquid Interface production. Yamamoto M, editor. PLoS One. 2016;11(9):1-17

79. Lan PX, Lee JW, Seol Y-J, Cho D-W. Development of 3D PPF/DEF scaffolds using micro-stereolithography and surface modification. Journal of Materials Science. Materials in Medicine. 2009;20(1):271-279

80. Brie J, Chartier T, Chaput C, Delage C, Pradeau B, Caire F, et al. A new custom made bioceramic implant for the repair of large and complex craniofacial bone defects. Journal of Cranio-Maxillo-Facial Surgery. 2013;41(5):403-407

81. Miyazaki T, Hotta Y, Kunii J, Kuriyama S, Tamaki Y. A review of dental CAD/CAM: Current status and future perspectives from 20 years of experience. Dental Materials Journal. 2009;28(1):44-56

82. Bhargav A, Sanjairaj V, Rosa V, Wen Feng L, Fuh JY. Applications of additive manufacturing in dentistry: A review. Journal of Biomedical Materials Research Part B: Applied Biomaterials. 2017

83. Dodziuk H. Applications of 3D printing in healthcare. Kardiochirurgia i Torakochirurgia Polska/Polish Journal of Thoracic and Cardiovascular Surgery. 2016;13(3):283-293

84. Chrzan R, Miechowicz S, Urbanik A, Markowska O, Kudasik T. Individually fitted hearing aid device manufactured using rapid prototyping based on ear CT a case report. The Neuroradiology Journal. 2009;22:209-214

85. Alhnan MA, Okwuosa TC, Sadia M, Wan K-W, Ahmed W, Arafat B. Emergence of 3D printed dosage forms: Opportunities and challenges. Pharmaceutical Research. 2016;33(8):1817-1832

86. Marro A, Bandukwala T, Mak W. Three-dimensional printing and medical imaging: A review of the methods and applications. Current Problems in Diagnostic Radiology. 2016;45(1):2-9

87. Chae MP, Rozen WM, McMenamin PG, Findlay MW, Spychal RT, Hunter-Smith DJ. Emerging applications of bedside 3D printing in plastic surgery. Frontiers in Surgery. 2015;2:25

88. Dhir V, Itoi T, Fockens P, Perez-Miranda M, Khashab MA, Seo DW, et al. Novel ex vivo model for hands-on teaching of and training in EUS-guided biliary drainage: Creation of "Mumbai EUS" stereolithography/3D printing bile duct prototype (with videos). Gastrointestinal Endoscopy. 2015;81(2):440-446

89. Neiman JAS, Raman R, Chan V, Rhoads MG, Raredon MSB, Velazquez JJ, et al. Photopatterning of hydrogel scaffolds coupled to filter materials using stereolithography for perfused 3D culture of hepatocytes. Biotechnology and Bioengineering. 2015;2(4):777-787

90. Mota C, Puppi D, Chiellini F, Chiellini E. Additive manufacturing techniques for the production of tissue engineering constructs. Journal of Tissue Engineering and Regenerative Medicine. 2015;9(3):174-190.

91. Shanjani Y, Pan CC, Elomaa L, Yang Y. A novel bioprinting method and system for forming hybrid tissue engineering constructs. Biofabrication. 2015 Dec 18;7(4):45008.

92. ProJet® 1200 Micro-SLA Printer [Internet]. [cited 2018 Mar 14]. Available from: http://www.cadventure.co.uk/wp-content/uploads/2017/06/3D-Systems_Micro-SLA_ProJet_1200_Tech_Specs_USEN_2016.12.01_a_WEB.pdf

93. 3D Stereolithography Printers [Internet]. [cited 2018 Mar 14]. Available from: https://www.3dsystems.com/sites/default/files/2017-05/3D-Systems_SLA_Specsheet_A4_US_2017.05.16_WEB.pdf

94. Technische Daten | Formlabs [Internet]. [cited 2018 Mar 14]. Available from: https://formlabs.com/de/3d-printers/technische-daten/.

95. Nobel 1.0A | SL 3D Printer | 3D Printer |XYZprinting [Internet]. [cited 2018 Mar 14]. Available from: https://www.xyzprinting.com/en-US/product/nobel-1-0a.

96. Micro Plus cDLM [Internet]. [cited 2018 Mar 14]. Available from: https://envisiontec.com/wp-content/uploads/2016/09/2017-Micro-Plus-cDLM.pdf

97. 3D printer – Wanhao [Internet]. [cited 2018 Mar 14]. Available from: http://www.wanhao3dprinter.com/

98. High Resolution 3D Printer – SLA DLP – Kudo3D Inc. [Internet]. [cited 2018 Mar 14]. Available from: https://www.kudo3d.com/

99. Liquid Crystal Precision – Photocentric Group [Internet]. [cited 2018 Mar 14]. Available from: https://photocentricgroup.com/lcprecision/?v=fa868488740a

100. Lee MP, Cooper GJT, Hinkley T, Gibson GM, Padgett MJ, Cronin L. Development of a 3D printer using scanning projection stereolithography. Scientific Reports. 2015 Sep 23;5(1):9875

101. Moran BD. Large area projection micro stereolithography. 2015. US14688187

102. Zheng X, Smith W, Jackson J, Moran B, Cui H, Chen D, et al. Multiscale metallic metamaterials. Nature Materials. 2016 Oct 18;15(10):1100-1106.

103. Zhou C, Chen Y, Yang Z, Khoshnevis B, Epstein DJ. Digital material fabrication using mask-image-projection- based stereolithography. Rapid Prototyping Journal. 2013;19(3):153-165.

104. Choi J-W, Kim H-C, Wicker R. Multi-material stereolithography. Journal of Materials Processing Technology. 2011 Mar 1;211(3):318-328.

105. Arcaute K, Mann B, Wicker R. Stereolithography of spatially controlled multi-material bioactive poly(ethylene glycol) scaffolds. Acta Biomaterialia. 2010 Mar 1;6(3):1047-1054.

CHAPTER-2

HIGH-RESOLUTION ELECTRIC-FIELD-DRIVEN JET 3D PRINTING AND APPLICATIONS

1. INTRODUCTION

The manufacturing technology for multi-material and multi-scale structure has been regarded as one of the most urgent needs for the manufacturing science. The development of new functional materials (functional gradient materials, biomimetic materials, and intelligent materials), tissue engineering (tissue scaffold, and organ), and next generation electronic products (embedded electronic products, flexible electronics, and wearable devices) has been proposing wide scientific and practical engineering requirements for the multi-material and multi-scale manufacturing technique. However, the existing manufacturing technologies are facing enormous challenges in the field of multi-material and multi-scale structure manufacturing [1, 2, 3].

3D printing has shown great potential and broad engineering application prospect in the field of multi-material and multi-scale 3D structure manufacturing due to its advantages including (1) the shape complexity; (2) the material complexity; (3) the function complexity and so on. In recent years, in order to achieve the goal of functional driven integrated manufacturing of "material-structure-devices," 3D printing technology has been developed from traditional controlling shape to controlling properties, from printing single homogeneous material to printing multi-material or functional gradient material, from macro-scale to micro-scale or multi-scale including macro/micro/nanoscale. The functional structure electronics, as one kind of typical multi-material and multi-scale 3D printing product, which is composed of structural materials (plastics, polymers, ceramics, metals, etc.), conductive materials (the conductive silver paste, the nanosilver conductive ink, etc.), dielectric materials (various insulating materials), and semiconductor materials. The feature size of printed structure includes multi-scale (macro/micro/nanoscale) [4, 5, 6, 7, 8, 9, 10]. However, most of the existing 3D printing technologies are limited to print single material, and it is difficult to achieve the multi-scale (macro/micro/nanoscale) heterogeneous 3D structure manufacturing.

Compared with other 3D printing processes, the 3D printing technologies (inkjet printing and electrohydrodynamic jet printing) based on materials' jet deposition have shown outstanding advantages for multi-material and multi-scale structure manufacturing. However, the inkjet printing is limited by the viscosity of printing material, which is usually less than 30cp. And the resolution of the current printed patterns is generally not more than 20 μm of line width. The electrohydrodynamic (EHD) jet printing [11, 12], which has been proposed and developed by Park and Rogers, has much higher resolution because of the Taylor cone induced by electric field. However, there are still many shortcomings and limitations as follows: (1) The height limitation of the printed structure, it is also difficult to achieve macro/micro scale integrated printing due to the limitation of 3 mm for the distance between nozzle and substrate; (2) The insulating substrate and its thickness limitation [13]. (3) The big challenging for the conformal printing on the surface of an existing object, especially on the inclined and curved surfaces [14].

In order to solve the problems of complex 3D structure manufacturing with multi-material and multi-scale, and to achieve the integrated manufacturing with heterogeneous multi-material in macro/micro multi-scale. A novel high-resolution 3D printing, named as high-resolution electric-field-driven jet 3D printing, which is based on the induced electric field and EHD cone-jetting behavior, has been developed to provide a feasible approach to implement additive manufacturing with multi-scale and multi-material at low cost. This chapter will introduce the principle, the simulation, the experiments, and the applications of EFD jet deposition 3D printing technology.

2. THE PRINCIPLE OF EFD JET DEPOSITION 3D PRINTING

2.1. The basic principle

The electric-field-driven (EFD) jet 3D printing system is mainly composed of a printhead, a three-axis translation stage, a high-voltage power supply, a signal generator, and a material feeding unit. Figure 1 shows the setup schematic of the EFD jet deposition 3D printing system. The nozzle of the printhead is directly connected to the anode of the high-voltage pulse power supply for inducing an electric field. The three-axis translation stage can be programmed to provide the planar movement of the target substrate for printing patterns and up-and-down movement of the nozzle for changing the intensity of the induced electric filed. Liquid materials are delivered to the printhead by the material feeding unit, which consists of a pneumatic control system and a material reservoir. The pneumatic control system can adjust the flow rate to optimize the shape of the pendent meniscus at the nozzle by changing the gas pressure. A signal generator is used to produce the modulated voltage command signal with desired frequency and duration to trigger the output of high-voltage power supply. The printing process can be recorded by a CCD camera.

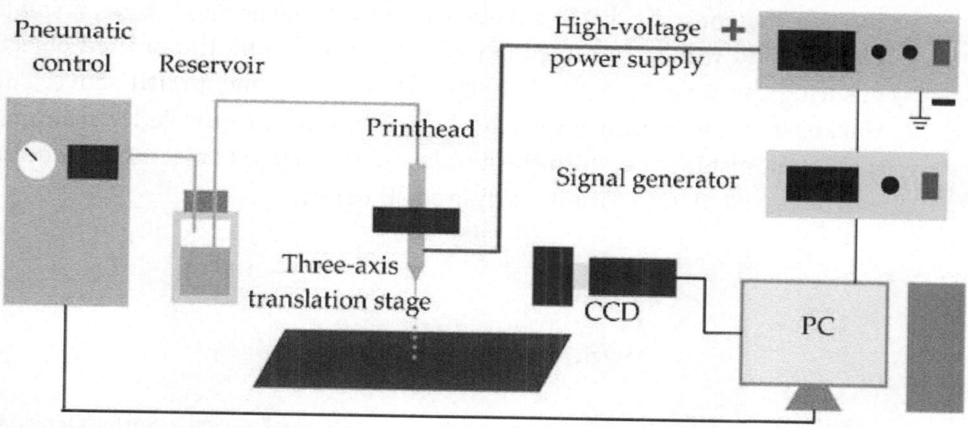

Figure 1. Schematic of system setup for the EFD jet deposition 3D printing.

Differing from the traditional pressure-driven 3D printing and the existing EHD jet printing, the proposed technique is a liquid ejection and deposition printing based on the electrostatic induction and EHD cone-jetting. The fundamental principle of the EFD jet 3D printing is illustrated in Figure 2. The printing process can be described as follows:

- ☆ With the movement of the nozzle with a high electric potential to the target substrate, the interaction between the nozzle and the substrate will be stronger and stronger by the effect of electrostatic induction. Then, an induced electric field will be generated between the nozzle and the substrate, as shown in Figure 2(a).
- ☆ The pendent meniscus induced by the gas pressure at the nozzle will be affected by the electric field causing the accumulation of the positive charges on its surface. Under the coupling effect of the electric field force, surface

tension, viscous force, and gas pressure, the meniscus will be elongated gradually to form a Taylor-cone.

- ☆ When the tangential electric stress exceeds the surface tension during a single pulse, a droplet or fine jet from the apex of the cone will be produced. The droplet diameter and jet size are significantly smaller than nozzle size, indicating the EFD jet printing can overcome the resolution limitation from the nozzle size.
- ☆ The simultaneous control of the ejection and movement of the printhead or substrate allows precise deposition of materials on the substrate, forming the first layer of printing object.
- ☆ Regarding the first layer as the target substrate, then moving up the nozzle to keep the constant distance between the first layer and nozzle, a stable electric field will be reformed between the nozzle and the printed object, ensuring the stability and reliability of the process mentioned above. The thickness of the printing layer must be appropriately controlled to maintain a constant distance between the nozzle and the printed object. This process will be repeated until forming a desired 3D object.

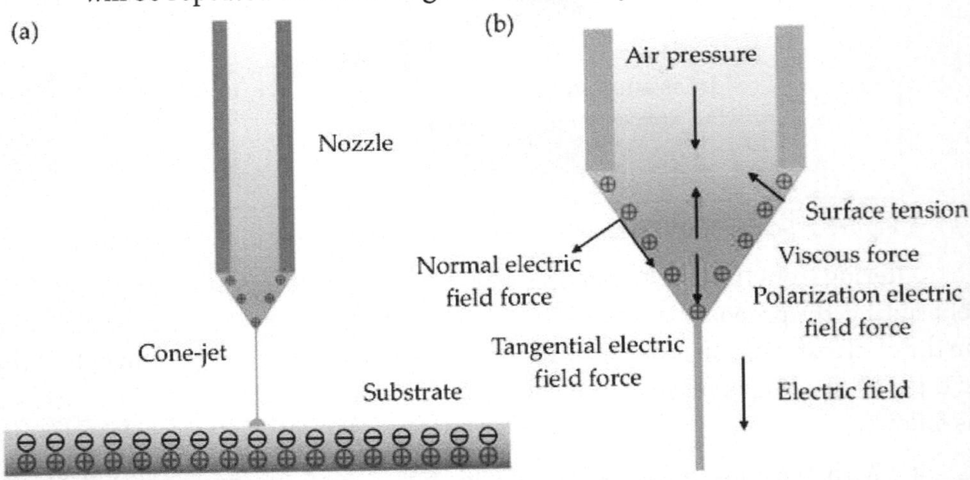

Figure 2. Schematic principle of the EFD jet deposition 3D printing: (a) the electrostatic induction between the nozzle and substrate; (b) stresses acting on the meniscus.

2.2. The two working modes

Unlike the traditional macro-scale 3D printing and micro-scale 3D printing, the macro/micro-scale 3D printing should take into account both the printing accuracy and the printing efficiency in the printing process [15, 16]. To ensure the capability of integrated printing of multi-scale complex 3D structure, and to better meet the requirements of the practical engineering of 3D printing. We proposed two kinds of working modes for this new technique, including the pulsed cone-jet mode and continuous cone-jet mode, to achieve both high precision and high efficiency during the printing process, as shown in Figure 3.

☆ **In the pulsed cone**-jet mode, the printing is a drop-on-demand process based on a pulsed voltage. When a pulsed voltage is applied, the meniscus will be deformed into Taylor-cone, and a micro-size jet is extracted to produce a micro-droplet. Then, the meniscus will be returned to the original position to produce a consistent and repeatable jet at the next pulse voltage. The printing material is deposited onto the substrate or the printed objects in the form of dots. The size of the printed dots is determined by process parameters involving the applied voltage, pulse duration time, and gas pressure, etc. The dots' spacing is mainly determined by the pulse interval time and the moving speed of the X-Y stage.

☆ **In the continuous cone**-jet mode, the DC-voltage is applied to drive and maintain the ejection behavior. The continuous liquid with a constant flow rate is rapidly pulled from the nozzle by the electric field force. Therefore, the printing material can be deposited on the substrate or the printed object in the form of lines (similar to FDM, but the dominant driving force is still the electric field force). Moreover, the line width is related to process parameters such as the voltage, the gas pressure, and the stage velocity, etc. The continuous cone-jet mode, based on continuous printing, possesses the ability of manufacturing large-area patterns and macro/micro-scale structures compared to the pulsed cone-jet mode.

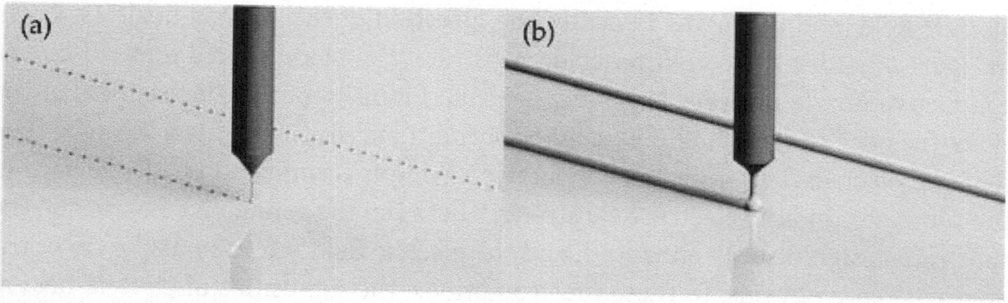

Figure 3. Two working modes (a) pulsed cone-jet mode; (b) continuous cone-jet mode.

Therefore, the proposed EFD jet deposition 3D printing possesses the following advantages:

☆ Suitable for both conductive and non-conductive (or insulating) substrates;

☆ Breaking through the height limitation of printed structures of traditional EHD jet printing;

- ☆ It can be widely used for various printing materials with large range of viscosity; and
- ☆ It also can be used to print on the uneven surface and the conformal surface.

3. THE NUMERICAL SIMULATION AND EXPERIMENTS VERIFICATION

In order to further reveal the injection molding mechanism and influencing rules of the processing parameters for EFD jet deposition 3D printing technology, this part will be focused on the investigation of printing performance under different conditions by means of numerical simulation and experimental verification. For the numerical simulation, the properties of substrate (or the printed layers), the printing height, and the variety of printing materials have been studied by the finite element simulation software (Comsol Multiphysics), and the simulation results are verified by experiments. The results can be seen as follows.

3.1. The effects of different substrate materials

In order to display the electrostatic induction between the nozzle and the substrate, the intensity and the distribution of the electric field around nozzles under four different kinds conditions (no substrate, PET substrate, glass substrate, and copper substrate) were studied by finite element simulation. The relative permittivity of PET, glass, and copper substrate is 4, 10, and 1, respectively.

The simulation results of distribution and intensity of electric field with different substrates were shown in Figure 4. The color represents the intensity of electric field, and the arrow indicates the direction of the electric field. As shown in Figure 4, under the condition of no substrate, the electric field emitted by the conductive nozzle diverges anywhere, and the intensity of the electric field at the tip of the nozzle is small. While under the conditions with substrates, it can be seen that the electric field from the conductive nozzle is terminated on the surface of the substrate, and the electric field strength between the conductive nozzle and the substrate is significantly enhanced, and the electric field intensity at the tip of the nozzle is more obvious. In these four conditions, the intensities of electric field at the tip of the nozzle are 618.1, 2794.8, 2794.9 and 3227.8 V/mm, respectively. The intensity of electric field on the copper substrate is significantly higher than that of the other three cases.

The reason why the substrate can change the electric field distribution is that the charges inside the substrate will redistribute by the effect of the electric field. For conductive substrate, there are lots of free charges inside the conductor, which can move freely inside the conductor. While for the insulating substrate such as PET and glass, only small amount of movable free charges can be found inside the insulating substrate, and most of charges are bounded in micro-displacement. Therefore, the electrostatic induction of the dielectric substrate is weaker than that of conductive substrate. When the conductive substrate stays in the condition of electrostatic balance, a large amount of free charges are gathered on the surface of the conductive substrate, which makes the direction of the electric field from the nozzle change to the conductive substrate. Therefore, a stronger electric field can be obtained between the nozzle and conductive substrate to provide a greater driving force for droplet jetting.

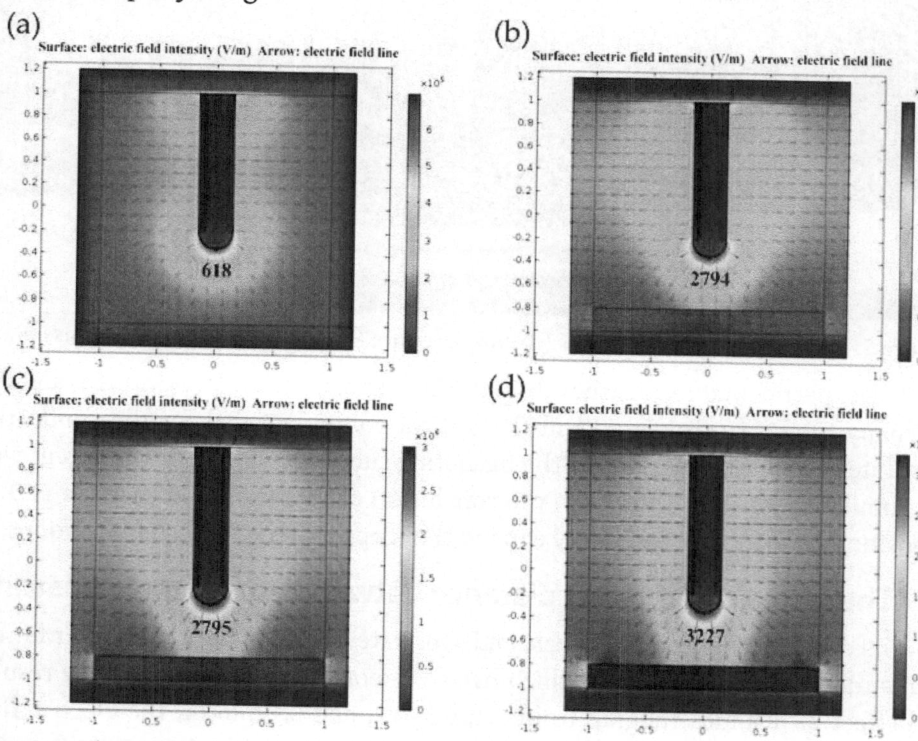

Figure 4. The simulation results of electric field under different substrate conditions: (a) no substrate, (b) PET, (c) glass, and (d) copper.

Compared with the existing EHD jet printing, the EFD jet deposition 3D printing has better applicability for different substrate materials. The EHD jet printing technology usually requires good conductivity of the target substrate, otherwise the insulating substrate with a thickness limitation must be placed on a grounding conductor. The EFD jet deposition 3D printing technology can be used in any material of substrates because the electrostatic induction can be generated in various materials, which is not limited by the conductivity or dielectric properties of the substrate. Figure 5 shows that the Taylor cone and stable cone jet can be formed at the nozzle by using for different material substrates (conductive stainless steel, semiconductor silicon chip, and insulating glass). The distance between the tip of the nozzle and the substrate is set as 2 mm, and the printing material is low viscosity resin (100 mPa.s). The critical voltage of cone jet for conductive stainless steel, semiconductor silicon chip, and insulator glass are 2100, 2500, and 3000 V.

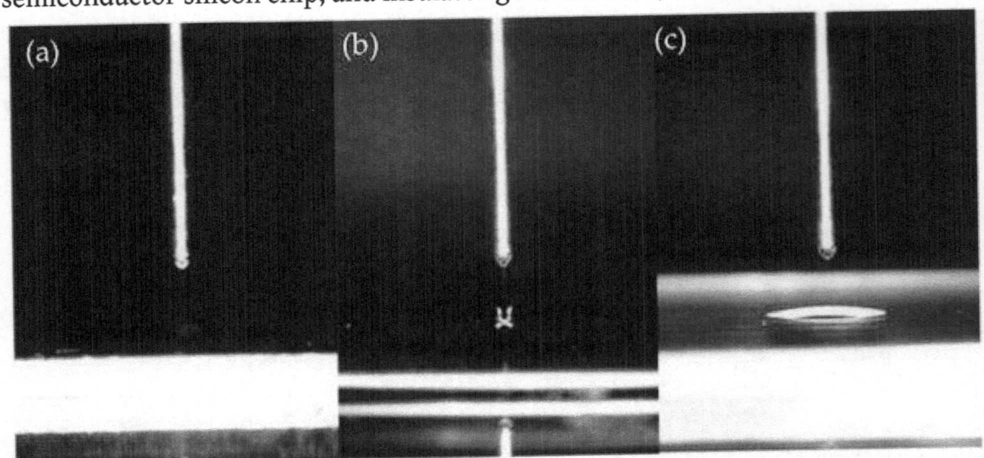

Figure 5. Different target substrates (a) copper plate; (b) silicon plate; and (c) glass plate.

The experimental results show that the EFD jet deposition 3D printing is suitable for many types of substrates. And the critical voltage needed for the conductive substrate is smaller than that for the insulating substrates, which is consistent with the simulation results. Therefore, the conclusion can be proposed that the EFD jet deposition 3D printing can greatly expand the scope and the field of applications.

3.2. The effect of different distance between nozzle and substrate

The distance between the nozzle and substrate is another key parameter for the EFD jet deposition 3D printing, which has an important impact on printing results. The effect of distance ranging from 0.4, 0.3, 0.2, and 0.1 mm on the electric field were simulated by COMSOL software, as shown in Figure 6. The intensity of electric field on the droplet surface is 3376.8, 3816.3, 4669.5 and 6910.5 V/mm, respectively. It can be seen that with the decrease of distance, the electric field formed between the conductive nozzle and the substrate is much enhanced, especially at the tip of the nozzle.

Due to the distance limitation between the conductive nozzle and the conductive substrate (or conductor under insulating substrate), the maximum distance of the EHD jet printing technology is usually not more than 3 mm, This caused great challenges in the macro/micro scale manufacturing for EHD jet printing. The EFD jet deposition 3D printing technology has broken through the distance limitation of the traditional EHD jet printing, and can truly realize the macro/micro-scale manufacturing. When the high voltage power is applied to the conductive nozzle, the required electric field can be formed between them under the action of electrostatic induction only if moving nozzle closed to the target substrate (or the printed layers). Therefore, this proposed technology can print on any structure surface.

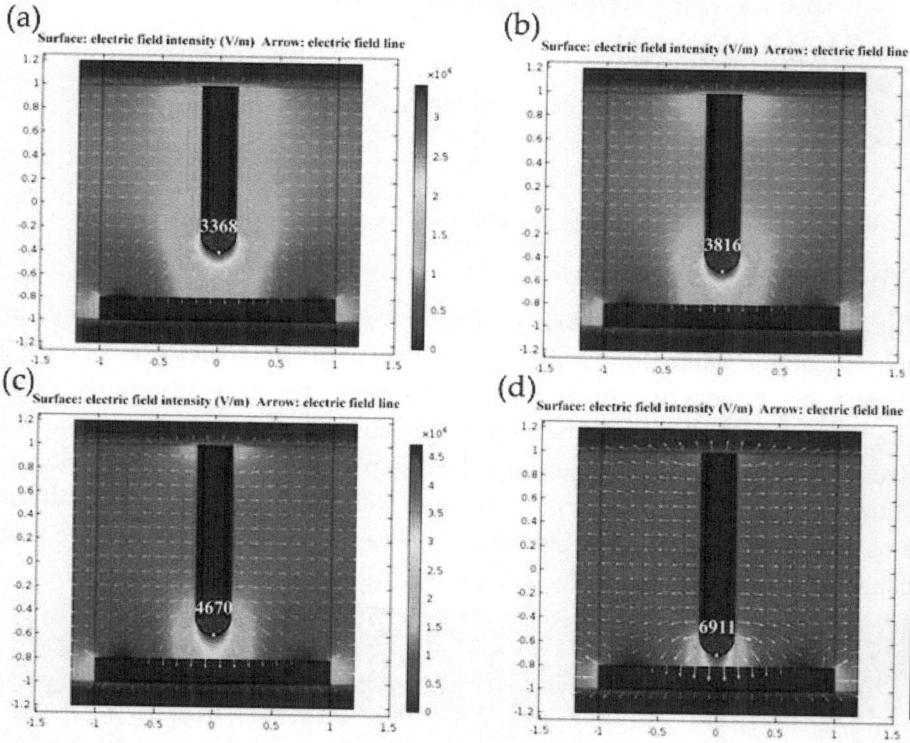

Figure 6. The simulation results of electric field at different nozzle heights: (a) 0.4 mm, (b) 0.3 mm, (c) 0.2 mm, and (d) 0.1 mm.

The experiments have been done to investigate the stability of the Taylor cone shape and the cone jet in the printing process influenced by the substrate thickness, as shown in Figure 7. During the experiments, the distance between the conductive nozzle and the glass substrate is set as 2 mm, and the printing material is low viscosity resin (viscosity 100 mPa.s), and the voltage is 3000 V. It can be observed that the thickness of glass substrate ranging from 1.2 to 12 mm, the EFD jet deposition 3D printing can achieve stable and reliable printing. The experimental results show that the EFD jet deposition 3D printing can truly achieve the 3D printing by means of electrostatic induction between the nozzle and the target substrate (or the printed layers) (Figure 7).

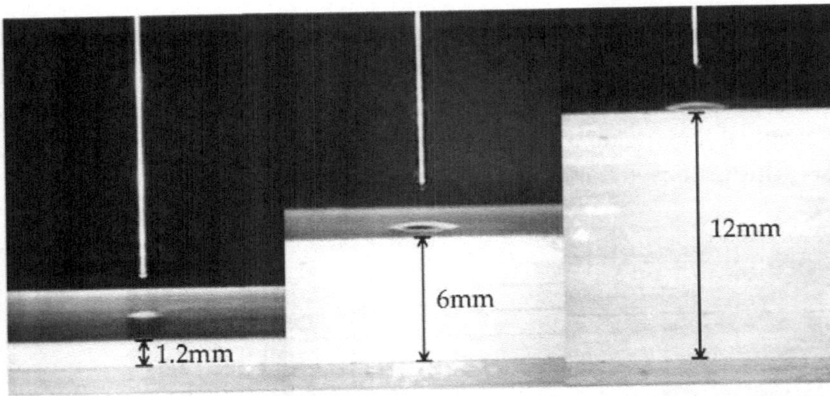

Figure 7. Reliably printing in variable heights and locations of printhead.

3.3. The printing materials

The available materials for EFD jet deposition 3D printing are extensive and variable, such as organic polymer materials, bio-materials, nanoscale composites, metal, and non-conductive materials. Furthermore, the viscosity range of the printing materials is very broad because of the enough electric stress provided by the strong electric field to drive jetting. The line patterns with high resolution and high quality (line-width roughness) have been well produced using four different types of materials, including the photosensitive resin, nanosilver conductive paste, polycaprolactone (PCL), and conductive silver adhesive, as shown in Figure 8.

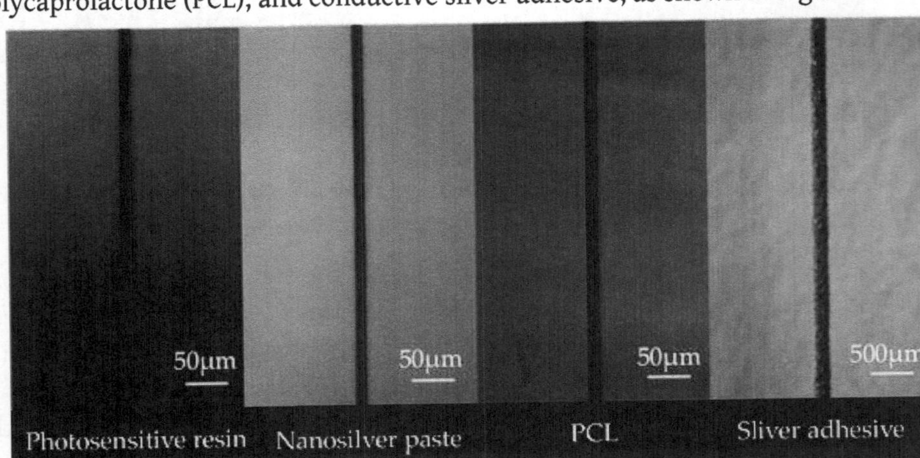

Figure 8. The features printed with various printing materials.

The materials in the experiment have different fluidic properties and the viscosity. The viscosity of the photosensitive resin is 800 cP while the viscosity of nanosilver conductive paste is 5000 cP. The heating system was used for molding of PCL because of its solid-state at RT. The nozzle with an inner diameter of 250 μm is adopted to print objects with a line width of 10 μm, where the reduction ratio in dimensions between the nozzle and the printed line reaches 25:1.

The viscosity of the conductive silver adhesives at ambient temperature is 8000 cP, the inner diameter of the nozzle is 250 μm, the line width of the object to be printed is 200 μm, and the print patterns must have good morphology. The results showed that the EFD jet 3D printing is suitable for almost any materials, compared to existing 3D printing technology. It is unexamined for its potential to provide high-resolution (that is, ~10um) patterning of materials with ultra-high viscosity.

4. THE APPLICATIONS OF EFD JET DEPOSITION 3D PRINTING

In this part, four typical printed structures were used to present the applications of EFD jet deposition 3D printing in multi-scale and multi material 3D printing: (1) lines and dots (one-dimensional structure); (2) high aspect ratio micro-scale "wall" structure (two-dimensional structure); (3) high resolution tissue engineering scaffold (3D structure); (4) 3D structure electronics (multi-scale and multi-material heterogeneous 3D structure).

4.1. Lines and dots (one-dimensional structure)

The droplet can be precisely deposited at the designated position by controlling the process of droplet jetting and the movement of stage. A dot array with the resolution of 50-60 μm and the different dot spacing has been printed to show the capability of drop-on-demand printing in the pulsed cone-jet mode, as shown in Figure 9(a). The dots' spacing decreases correspondingly with the decreasing of stage moving speed. Using the continuous cone-jet mode, a line pattern has been successfully printed on a glass slide, as shown in Figure 9(b). The average line width in this pattern is about 40 μm, and the line pitch is about 150 μm. Due to the ability of depositing materials directly as desired patterns on the substrate with a simple fabrication process and high efficiency, the proposed printing method can be adapted for applications in thin-film transistors, optical elements, organic light-emitting diodes, photonics crystals, and DNA microarrays.

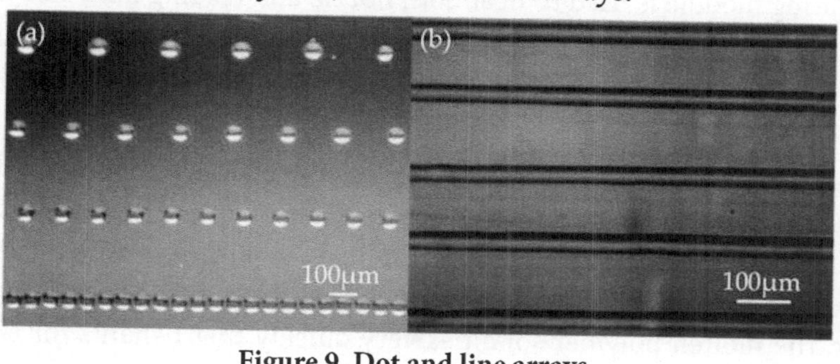

Figure 9. Dot and line arrays.

By using the conductive nanosilver paste, the proposed printing method can be applied for fabrication of metal-mesh patterns used in various electronics such as flexible displays, solar cells, touch panels, etc. Figure 10 shows the optical images of the printed metal-mesh patterns. The line width and spacing of the metal mesh in Figure 10(a) are 20 and 250 μm, respectively. And the line width and spacing in Figure 10(b) are 10 and 150 μm, respectively. The line resolution of the printed metal-mesh patterns is less than 20 μm, which is almost invisible to the naked eye. It indicates that the proposed technology is promising to fabricate an invisible fine transparent electrode with good electricity and optical properties, which can be widely applied to electronic devices without any cosmetic issues due to the appearance of metal pattern.

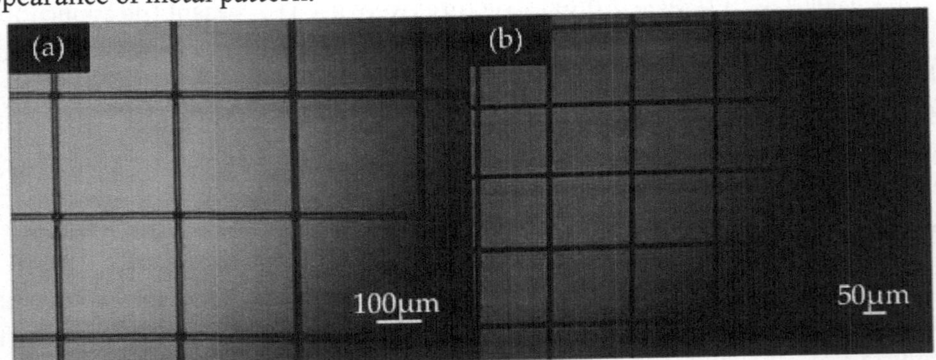

Figure 10. Metal-mesh patterns with (a) line width of 20 μm and spacing of 250 μm; (b) line width of 10 μm and spacing of 150 μm.

4.2. The micro-scale "wall" structure (two-dimensional structure)

The EFD jet deposition 3D printing technology is mainly used for liquid printing materials, it also can be used for printing molten polymer materials by changing the nozzle structure. The material feeding unit is integrated into the printhead to shorten the distance between material feeding unit and the nozzle, because solid state printing material is difficult to be delivered to the nozzle through pipeline. The double heating module is used to heat both nozzle and feeding unit. The purpose of heating feeding unit is to keep the printed material in the melting state with certain fluidity, and that of heating nozzle is to ensure the quality and precision of the printing process.

Instead of utilizing polymer solutions as the printing material, the molten EFD jet 3D printing employs molten polymers as the printing materials. Due to the printing material is PCL with a melting point of about 60°C. A heating module with a heating temperature of 80°C is utilized to melt the solid polymer into flowing melts. Moreover, the molten polymer solidifies very quickly that benefits for the layered manufacturing of high aspect ratio structures. Figure 11shows a micro-scale high aspect ratio wall structured cylinder with diameter of 20 mm, the wall thickness of 20 μm, and the height of 550 μm (continuous stacking of 20 layers). Compared to the traditional EHD printing, The experimental results show the molten EFD jet 3D printing offers a promising approach to produce the micro/nanostructures with ultra-high aspect ratio at low cost and high throughput.

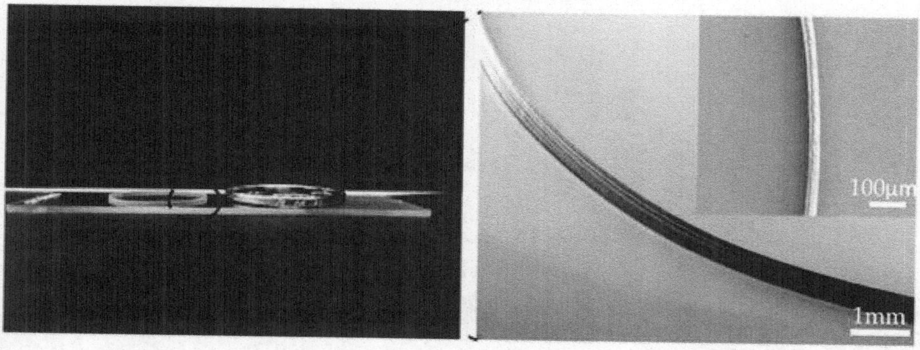

Figure 11. A cylinder structure with diameter of 20 mm, height of 550 μm, and wall thickness of 20 μm.

4.3. Tissue engineering scaffold (3D structure)

A typical macro/micro-scale tissue engineering scaffold is a 3D porous structure for transporting nourishment and excreting metabolites for cell growth. A desirable scaffold is characterized by controllable porosity, pore size, and pore distribution, which can provide cells with sufficient oxygen and nutrient supply. It is challenging for the current manufacturing technologies to the fully controlled orderly morphology and accuracy as requested. Polycaprolactone (PCL) with good biodegradable performance serves as the printing material of the biological scaffold, a molten EFD jet 3D printing was employed to fabricate 3D scaffold, as shown in Figure 12. The printing process parameters are set as follows: voltage 2.8kV, pressure 20 kPa, moving speed 5 mm/s, and heating temperature of 80°C. The overall size of the scaffold is 4 mm x 4 mm, the line width is 60 μm, the period is 300 m, and the height is 300 μm, shown in Figure 13. The experimental results confirm that the EFD jet deposition 3D printing possesses a very prominent ability for the macro/micro scale printing.

Figure 12. The macro view of a tissue engineering scaffold.

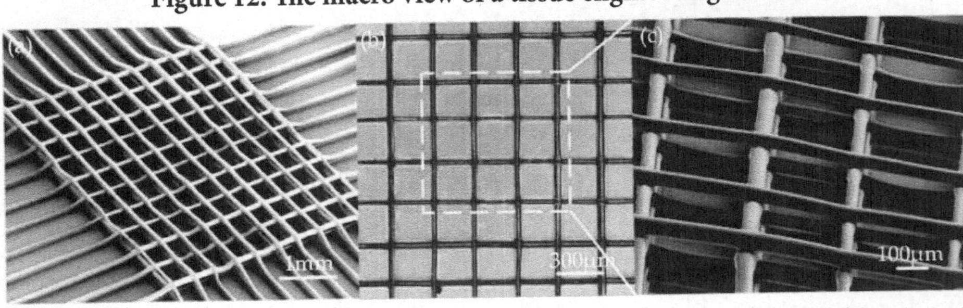

Figure 13. A tissue engineering scaffold: (a) overall view; (b) top view; and (c) microview.

4.4. 3D structural electronic (layered heterogeneous structure)

3D structure electronic is a typical multi-material structure product, which is widely used in many fields, such as aerospace, national defense, biological medicine and so on. However, how to realize the manufacturing of 3D structure electronic products with high efficiency and low cost is a very huge challenging problem. The EFD jet deposition 3D printing with multi-nozzle can provide an effective method for the manufacturing 3D electronic. In this case, two kinds of printing materials, photosensitive resin (viscosity 800 mPa s) and conductive silver (viscosity 8000 mPa s), are used. The photosensitive resin is used to construct main body of the 3D structure electronics, while the conductive silver paste is used to print interconnecting wires. The processing parameters of photosensitive resin are printed with voltage 3.2 kV, gas pressure 50 kPa, and moving speed 30 mm/s, while the printing process parameters of conductive silver paste are voltage 2.0 kV, gas pressure 20 kPa, and moving speed 3 mm/s.

As shown in Figure 14, the printed structure made of photosensitive resin is a circular platform with a diameter of 7.5 mm at the bottom surface, a diameter of

5.5 mm at the top surface, and a height of 4 mm. There is a printed conductive wire made of conductive silver paste on the top surface of the round platform. Conductive silver paste can be cured at room temperature without heating and other post-processing, which will never damage printed main body of 3D structure electronics.

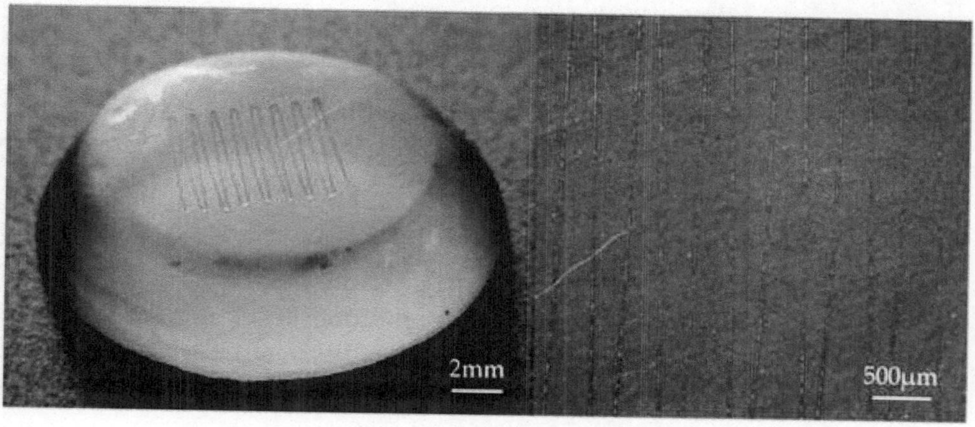

Figure 14. A 3D structure electronic device.

Therefore, combining the capability of printing variable materials and the use of multi-nozzle technology, the EFD jet deposition 3D printing presents a very prominent advantage and great potential in the multi-scale and multi-material 3D printing. This technology provides a new solution for the integrated printing of heterogeneous multi-material, multi-scale (macro/micro-scale) structure.

5. CONCLUSION

A novel high-resolution 3D printing, named as high-resolution Electric-field-driven Jet 3D Printing, which is based on the induced electric field and EHD cone-jetting behavior, has been developed to provide a feasible approach to implement additive manufacturing with multi-scale and multi-material. In this chapter, we introduced the system setup and the principle of EFD jet deposition 3D printing. Then, the jetting molding mechanism and influencing rules of operating parameters were revealed by the numerical simulation and verified by the experiments. Finally, four kinds of typical applications including transparent conducting electrodes, high aspect ratio "wall" structure, tissue engineering scaffold, 3D electronics, etc., are presented in detailed. As a result, this new technology offers a novel solution for fulfilling multi-scale and multi-material 3D printing at low cost and good universality as well as high resolution.

REFERENCES

1. Wegst U, Bai H, Saiz E, et al. Bioinspired structural materials. Nature Materials. 2015;14:23-36. DOI: 10.1038/nmat4089

2. Rosen D, Gibson I, Stucker B. Additive Manufacturing Technologies. New York: Springer; 2010. DOI: 10.1007/978-1-4939-2113-3

3. Campbell J, McGuinness I, Wirz H, et al. Multimaterial and multiscale three-dimensional bioprinter. Transactions of the ASME Journal of Nanotechnology in Engineering and Medicine. 2015;6:021005. DOI: 10.1115/1.4031230

4. MacDonald E, Wicker R. Multiprocess 3D printing for increasing component functionality. Science. 2016;353(6307):aaf2093. DOI: 10.1126/science.aaf2093

5. Ota H, Emaminejad S, Gao Y, et al. Application of 3D printing for smart objects with embedded electronic sensors and systems. Advanced Materials Technologies. 2016;1(1):1-8. DOI: 10.1002/admt.201600013

6. Truby R, Lewis J. Printing soft matter in three dimensions. Nature. 2016;540:371-378. DOI: 10.1038/nature21003

7. Bartlett N, Tolley M, Overvelde J, et al. A 3D-printed, functionally graded soft robot powered by combustion. Science. 2015;349(6244):161-165. DOI: 10.1126/science.aab0129

8. Wehner M, Truby R, Fitzgerald D, et al. An integrated design and fabrication strategy for entirely soft, autonomous robots. Nature. 2016;536:451-466. DOI: 10.1038/nature19100

9. Lewis J, Ahn B. Device fabrication: Three-dimensional printed electronics. Nature. 2015;518:42-43. DOI: 10.1038/518042a

10. Kong YL, Tamargo IA, Kim H, et al. 3D printed quantum dot light-emitting diodes. Nano Letters. 2014;14(12):7017-7023. DOI: 10.1021/nl5033292

11. Raje PV, Murmu NC. A review on electroohydrodynamic-inkjet printing technology. International Journal of Emerging Technology and Advanced Engineering. 2014;4:174-183

12. Huang Y, Bu N, Duan Y, et al. Electrohydrodynamic direct-writing. Nanoscale. 2013;5:12007-12017. DOI: 10.1039/C3NR04329K

13. Rahman K, Ali K, Muhammad NM, et al. Fine resolution drop-on-demand electrohydrodynamic patterning of conductive silver tracks on glass substrate. Applied Physics A. 2013;111(2):593-600. DOI: 10.1007/s00339-012-7267-x

14. Tse L, Barton K. A field shaping printhead for high-resolution electrohydrodynamic jet printing onto non-conductive and uneven surfaces. Applied Physics Letters. 2014;104(14):143510-143514. DOI: 10.1063/1.4871103

15. Phung TH, Kim S, Kwon KS. A high speed electrohydrodynamic (EHD) jet printingmethod for line printing. Journal of Micromechanics and Microengineering. 2017;27(9):095003(8pp). DOI: 10.1088/1361-6439/aa7c6b

16. Mishra S, Barton KL, Alleyne AG, et al. High-speed and drop-on-demand printing with a pulsed electrohydrodynamic jet. Journal of Micromechanics and Microengineering. 2010;20(20):095026. DOI: 10.1088/0960-1317/20/9/095026

CHAPTER-3

THE EVOLUTION OF 3D PRINTING IN AEC: FROM EXPERIMENTAL TO CONSOLIDATED TECHNIQUES

1. INTRODUCTION

3D printing can be nowadays considered a consolidated technology, at least in its technical aspects. However, the adoption of such manufacturing technique to architecture engineering and construction (AEC) is not widespread yet, as the sector is not yet completely ready for the introduction of innovative production methods, in comparison to other more innovative sectors. Some experimental case studies have been developed looking at possible applications of 3D printing in architecture and construction, but the gap to close is now related to a consolidated way of employing innovative manufacturing techniques.

2. PRODUCTION TECHNIQUES IN AEC

The historical evolution of architecture is closely linked to that of construction techniques. The combination of available techniques and workforce—in quantity and quality—has driven the sector since antiquity, and architects had to know and carefully consider them as a premise of their design. Moreover, while some techniques have emerged from within the field of architecture, in the effort of solving construction problems, very often it was the spillover of advancements in other fields of science and technology that determined the adoption of new production techniques in architecture.

While such combination of workforce skills and production techniques has been consistent throughout the centuries, there have been some radical paradigm changes in their combination. In particular, while a sort of batch production of some architectural elements was present since antiquity, as well as in gothic architecture—as for bricks, tiles, and column drums—starting with the industrial revolution, such production in series acquired a more industrial character, and the relevance of skilled labor started to decline, while mechanized processes took off as the most decisive factor in production costs and quality. Modularity, which previously was rather an ideal set of geometrical relationships and proportions related to orders, started to become a necessary way of streamlining the production in series of identical base components, the only way industrialization could lower production costs as well as assembly times and efforts. Architectural practices and theories had to reflect these needs, and especially with the Modern movement, the trend toward simplification and use of standardized elements became common practice. The case of 'The Eames House, Case Study House 8' by Charles and Ray is a paradigmatic example thereof: the building was even designed and assembled starting from 'off-the-shelf' standard pieces, while trying to create an individual architectural character. Production in this case was a given before the design, and not the result thereof. Fast forwarding in history, the use of building information modeling (BIM) software has connected this trend to the realm of the design in the virtual (software) environment, especially as it allows and even encourages the use of available and industrially pre-fabricated architectural elements, such as doors and windows, and also rebars, trusses, and the like.

On the other hand, starting from the 1960s, the degree of geometric freedom and control over the produced elements started to increase through the use of computer-aided design (CAD) and computer-aided manufacturing (CAM) software, even though the constraints of a required standardization of elements continued to be present for a cost-effective production. Through Bézier curves and Non-Uniform Rational B-Spline (NURBS) modelers, it was now possible to create more organic and complex shapes. An early example thereof was the Renault 'Unisurf' software used to design and produce car parts. However, the process often required non-computer-controlled phases and the mass-production of standardized pieces.

It was in the last 10–20 years that a more streamlined and integrated use of computer numerical controlled (CNC) machines started to allow for a new change in paradigm within the architectural field. While a few centuries ago, the spillover of industrialization techniques meant that standardization and simplification had to become the design approach to architectural projects because industrial production required identical elements to be mass-produced in order to lower the cost per unit, now it became possible to cost-effectively mass-produce elements that are different from one another, i.e., customized elements. It is the 'mass-customization' paradigm. The use of CAD/CAM software became the key tool in the hands of designers and architects to harvest this new production potential. In fact, the 'virtual' design within the software could be now transformed into something tangible driving the production machines directly from the computer and without the need for any 'translator' or skilled human intermediary. As in the First Industrial Revolution, workforce manual skills were not relevant anymore, but unlike under the previous paradigm, it was now not necessary or advantageous to reduce the complexity of the design elements and to embrace radical simplification. As we will see dealing with 3D printing techniques, it is worth noticing that this new approach started off as a convenient tool for fast prototyping, but due to technical advancements, it is potentially becoming a method for the production of final parts or even entire architectures, as it has already become a production technique in some fields of advanced engineering, such as aeronautics.

3. 3D PRINTING HISTORY RELATED TO CONSTRUCTION METHODS

Additive manufacturing (AM) is possibly the most disruptive production paradigm stemming from the adoption of CNC machines. It promises to transform a(ny) virtual shape designed in a software environment to a real-world object, as much as 2D printing is transforming virtual pixels into ink dots on a sheet. It requires that the object to be printed be 'rasterized' into discrete elements, which usually is performed through the use of Mesh geometries in the CAD environment. More often than not, additive manufacturing techniques are actually working by layered 'slices' (sections) of the desired object, so that the final shape results from the combination of subsequent, 2D designed, layers of material with a standard thickness.

3.1. History and evolution

1980–1981: Hideo Kodama (Nagoya Municipal Industrial Research Institute) invented and described two first additive manufacturing techniques based on photo-hardening of plastic polymers. This seminal work can be considered the ancestor of both photopolymerization and stereolithography. An application for patent was filed, but the inventor did not follow up within the required one-year deadline after application [1, 2].

1984: Jean-Claude André (CNRS), Alain le Méhauté (CGE/Alcatel) and Olivier de Witte (Cilas) filed an application for patent of stereolithography, i.e., an additive manufacturing method whereby a laser beam selectively hardens a UV-sensitive liquid resin, following a sequence of cross-sections of the object to be printed. The patent filing was abandoned, and Chuck Hull filed a patent, granted in 1986. The system was based on ultraviolet laser light beams hardening cross-section by cross-section a resin contained in a vat. The .stl file extension Hull adopted is still in use today for most AM. He also founded 3D Systems, a company manufacturing 3D printers.

1987: Carl R. Deckard invented at UT-Austin the selective laser sintering technique, based on high-power (usually pulsed) laser beam that selectively fuses powder particles along cross-sections of the desired shape. The powder can consist in plastic, metal, ceramic or glass, and is usually pre-heated in the bed just below the fusion point. A patent for a similar technique was filed in 1979 by R. F. Housholder, but it was not commercialized.

1989–1990: S. Scott Crump invented and patented the most popular 3D printing technique to date, especially for hobbyists and low-budget labs: fused deposition modeling (FDM). It consists in the deposition of fused material—most commonly plastic—layer by layer, according to a .stl file. The first machines were commercialized by Scott Crump's company Stratasys starting from 1992, and a patent was granted (expired in 2009).

1993: MIT developed what, strictly speaking, was considered 3D printing. The technique consisted in the binding—layer by layer—of a bed of powder using an inkjet printer, hence the name. In 1993, yet another technique was introduced by Sanders Prototype, Inc., now Solidscape: the 'dot-on-dot' technique. It was based on polymer jetting with soluble supports, yielding very high-precision results. The models were originally printed in wax.

1995: The Fraunhofer Institute ILT, Aachen, invented the selective laser melting process. The process—which yields precise and mechanically strong outputs, given the use of metal alloys, and can handle nested and intricate geometries—consists in the melting, layer by layer, of metal powder by means of a laser beam. Selective laser sintering is a similar process, whereby metal powder is not completely fused, hence does not form as much of a coherent and homogeneous mass as an output.

1999: Bioprinting techniques were successfully experimented at Wake Forest Institute for Regenerative Medicine.

2004: Adrian Bowyer developed the RepRap open-source project, aimed at creating self-replicable 3D printers, in an effort to diffuse and democratize AM technology.

2008: Shapeways was launched in the Netherlands. It consists in an on-line service, allowing users to send 3D files to have objects printed and sent to the required address. The service uses various techniques and materials, which today include several precious metals.

2009: Makerbot created a DIY kit for 3D printers which will highly contribute to the diffusion of the technique in many households.

2011: The opportunities offered by 3D printing techniques as production rather than pure prototyping tools were made even clearer by the Southampton University Laser Sintered Aircraft (SULSA), an unmanned aircraft whose structure was printed, from the wings to the integral control surfaces by a laser sintering machine, with a resolution of 100 micrometers per layer. The unmanned aerial vehicle (UAV) could be assembled without tools, using 'snap fit' techniques.

2014: Airbus Operation GmbH filed a patent for 3D printing an entire airplane structure. The technique is interesting also due to the 4D-printing-like features: a study on materials deformation, especially with respect to each other, is used to further strengthen the structure, by exploiting the resulting forces.

Overall, while the seminal ideas of the main additive manufacturing techniques dates back to the 1980s, further development and combination among techniques have gradually implied a shift in the potential use. In fact, while AM started as a means to rapid prototyping objects, especially for engineering—where the limited availability of materials and the lack of mechanical strength was not an issue—it now starts being adopted as a whole new way of producing final elements, given the improved quality of the output and the materials that can be used. Such opportunities could potentially disrupt the entire industrial processes and supply chain, enabling diffused fabrication facilities to such an extent that a so-called '0 Km factory' paradigm could emerge. 'Called microfactories, these diminutive factories drastically change how we produce large consumer goods for unique local needs' [3].

Finally, it is worth noting that, having these techniques not reached the full maturity phase yet, it would be pointless to analyze all the alternative methods and machines which have been invented and adopted for the most diverse projects. It seems therefore more useful analyzing the main categories, trying to provide a taxonomy thereof, bearing in mind that research is currently blossoming in the field, often hybridizing techniques to reach specific goals.

3.2. Taxonomy (and best sorting criteria): a material-state-driven categorization

Devising the 'most appropriate' classification criteria for additive manufacturing is not an easy task. Different approaches have in fact been taken into account in literature for classifying additive manufacturing processes. 'In particular Karunakaran exposes different possible options. A first option is to take as the driving aspect the type of material printed by the machines, which can turn out to be problematic because some machines can print more than one material typology. A second option refers to the material matrix, thus the ability of printers to work with a monolithic, composite, or gradient matrix, in terms of materic composition and properties, but it may result too specific with respect to the scope of the research. Another possible classification is according the final application of printed objects, which ranges from the visualization model to the high-end engineering part; again here, some printers may be used for different purposes, and moreover this subdivision would not clarify the different classes of layer manufacturing technologies and their behaviours. Always according to Karunakaran, more subdivision options can be referring to number of materials involved, on the energy source (laser, EB or arc) used, on the Boolean nature of the manufacture (laminated, powder-bed or deposition) or differencing methods of joining particles, but these approaches are too generic or too specific, not allowing a proper classification of the processes. The approach used by Gibson, is to manage the additive manufacturing techniques according to the starting condition of the material before it is worked by the machines. He defines liquid polymers, discrete particle, molten material and solid sheet systems. Often machines can print different classes of materials, and for different final purposes, but each printer can handle just materials in specific initial states, therefore this criteria is defining a proper subdivision which highlights the characteristics of the material processing, defining advantages and disadvantages of every process category' ([4], pp. 38–39).

We decided to adopt as sorting criteria two main aspects, which are *de facto* combined within the current digital fabrication techniques: 'state' of material, and additive process. In fact, groups of machines will tend to differ based on the state the materials come—here, we refer to grains, filaments, and liquid as 'states'—much more than they differ based on the kind of material. Many machines will be able to use different kinds of plastics and even metals, but require them to be in one specific state. As to the process, it tends to correlate strictly with the material state: for instance, any sintering technique requires a bed of grainy material, as it acts through the bonding of some grains as a way to create the final shape. Grains are a geometrical pre-requisite thereof. Similarly, stereolithography requires liquid resins to be shaped and hardened through light, which again requires a specific material state as a starting point. Based on the chosen sorting criteria, the main available additive manufacturing techniques can be summarized as follows.

3.2.1. Extrusion of fused/liquid material

These techniques share the common feature of a 'printing head' consisting in a moving nozzle that deposits layer upon layer of material.

3.2.1.1. Semisolid material extrusion

There are two main machine types: Gantry (or Cartesian) and Delta (Figure 1). Gantry is based on an extruder moving along the Cartesian X- and Y axes, while the plate is moving along the Z axis layer by layer. Delta systems, on the contrary, are based on three arms connected to universal joints at the base, which move within parallelograms, maintaining a lightweight end-effector in the right orientation. It yields faster and more accurate output, also given the lightweight traveling parts. While plastics are the main material used with this technique—specifically thermoplastics, especially ABS and PLA, and also nylon, PET, HIPS and TPU—ceramics, clay and cement were recently experimented with.

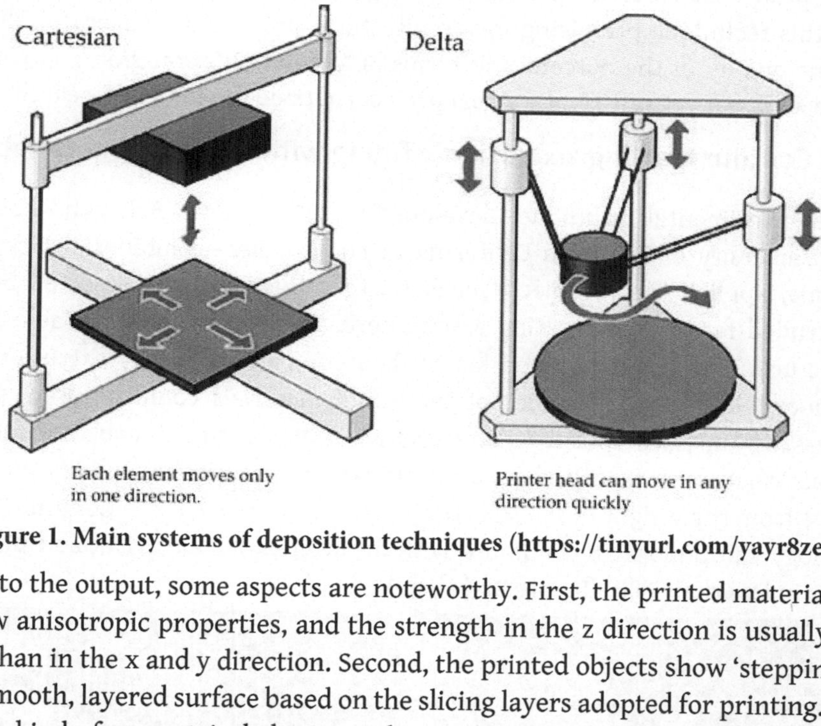

Figure 1. Main systems of deposition techniques (https://tinyurl.com/yayr8ze5).

As to the output, some aspects are noteworthy. First, the printed material tends to show anisotropic properties, and the strength in the z direction is usually much lower than in the x and y direction. Second, the printed objects show 'stepping', i.e., a nonsmooth, layered surface based on the slicing layers adopted for printing. Third, not any kind of geometrical shape can be produced with this technique: in fact, a maximum 45° of overhang, slanted parts can be produced without the creation of extra supports, which need to be later removed. Lastly, speed is a serious limitation for this technique to be used outside the boundaries of mere prototypation: a cube of 20 × 20 20 cm may require more than 24 h to be printed.

Other two less common machine types are the polar and the robotic arms. Polar machines work based on an angle and a length, and need only two motors, while the Gantry needs three. The robotic arm is not just a printing machine per se, but a printing head can be attached to a robotic arm. Potentially, it delivers much greater flexibility and printing dimension, especially if the arm is not fixed on the ground. Both techniques are quite experimental and not very widespread.

3.2.1.2. Semiliquid material extrusion

While not a consolidated technique, it has been experimented with by artists and researchers. It consists in the extension of the previously analyzed technique to the use of clay and similarly 'wet' materials. The main difference—though the process tends to vary for each experiment—is the absence of a heated print head, since the material does not have to be fused, while some kind of pressurized mechanism is usually present to force the muddy material through the nozzle. The possibility of using typical construction materials in architecture—such as clay and concrete—makes this technique promising for architectural projects. However, for the time being the quality of the outcomes in terms of 'resolution', precision and printable geometries is not yet sufficient for real projects outside the field of research.

3.2.1.3. Contour crafting: extrusion + filling with semifluid materials

The experimental technique—developed in 1998 by Prof. Behrokh Khoshnevis at the University of Southern California in Los Angeles—combines the extrusion technique, applied to the object 'surfaces', to a filler material injected between the extruded faces, thus creating a solid core. The technique is suitable for the architectural scale, as it is much faster than comparable purely extrusion-based techniques, while 'a wide choice of semi-fluid materials could be used, such as polymers, ceramics, composite wood materials, mortar, cement, concrete and other materials, that once deposited by a nozzle are able to quickly solidify and resist pressure from the weight of the structure itself. [...] Currently, the Contour Crafting technology can build a 185 m^2 house with all utilities for electrical and plumbing systems in less than 24 h' (Figure 2) ([4], p. 119) .

Figure 2. Contour crafting (https://tinyurl.com/y7xsbq83).

3.2.1.4. Concrete printing

Similar to contour crafting, developed at Loughborough University in the United Kingdom since 2004, it is similar to contour crafting, but allows to control the resolution of the nozzle for the deposition of both bulk materials and fine detail within the same process.

3.2.1.5. Metal extrusion (FDMm)

It encompasses a series of alternative experimental techniques that are either an adaptation of the semisolid material extrusion technique to low-melting-point metal, or the use of gas metal arc fusion welding robots.

3.2.2. Bonding of granular materials

Unlike the previously analyzed techniques, this series of techniques is based on the selective 'bonding' of grains of material previously disposed in an array. The advantage of this set of techniques lies in the almost infinite freedom of geometrical shapes it can produce, since no supports are needed and even nested shapes are printable in one step.

3.2.2.1. Binder jetting

The process consists in a multinozzle inkjet print head moving, layer by layer, on a 'powder bed', previously laid on the build platform. While a sweeper blade or roller evenly distributes the powder across the bed, the head selectively jets a binder solution, which solidifies the powder according to the section at stake. The bed is then lowered layer after layer. Different materials can be used as powder, including originally starch and gypsum plaster, while the binder—mostly water—can also contain dyes and other substances impacting on the physical properties of the powder (such as viscosity and surface tension). 'The resulting plaster parts typically lack "green strength" and require infiltration by melted wax, cyanoacrylate glue, epoxy, etc. before regular handling' ([4], pp. 53–55). The results of such technique tend to lack accuracy.

3.2.2.2. Selective laser sintering (SLS)

The process is generically called 'powder bed fusion', and it uses high-power laser to bond together the particles of material. Similar to the binder jetting process, the process consists in the selective hardening/binding of a powder bed. However, in this technique, the hardening happens through a laser beam that follows the cross section of the relevant layer. The material is heated just below the boiling point (proper 'sintering') or above it (selective laser melting). The process is completed layer by layer.

The main disadvantage of the process is the relatively high cost of the powerful lasers needed to print in materials other than composites, plastics and waxes, and the relatively weak mechanical performance of composite powders suitable for engineering applications. The advantages are numerous, ranging to the already mentioned geometrical freedom, to the fact it does not need much additional tooling after the object is printed. Moreover, the results can be very precise with high resolution (Figure 3).

Figure 3. Example of SLS production (https://tinyurl.com/y7bybghk).

3.2.2.3. Selective inhibition sintering (SIS)

This technique, developed by Dr. Behrokh Khoshnevis and his team at the University of Southern California, tries to address the trade-offs between the cost of high-power lasers for sintering metal, and the weak mechanical performance of composite materials suitable for lower-power lasers. 'In fact the principal innovation behind the SIS technique is the prevention of selected regions of each powder layer from sintering, achieved by operating on the regions external to the part in each layer with a "sintering inhibitor". A commercial piezoelectric print head is utilized to deposit a liquid chemical solution (inhibitor) at the periphery of the part for each layer. When all the layers have been treated, the entire part is removed from the machine and bulk sintered in a conventional sintering furnace. The inhibitor deposited at the part's boundary decomposes into hard particles that impede the sintering process. The particles in this region are prevented from fusing, allowing for removal of inhibited boundary sections and revealing of the completed part. It is easiest to think of the part as if it were encased in a sacrificial mold' ([4], p. 59).

This technique is still experimental, but is promising due to the lower costs implied by the use of conventional print heads available on the market, while it manages to produce full-metal parts with strong mechanical performances.

3.2.3. Photo polymerization of liquid materials

3.2.3.1. Stereo-lithography (SLA)

As seen, stereolithography was invented in the 1980s and consists in a technique whereby a laser beam selectively hardens a UV-sensitive liquid resin in a vat, following a sequence of cross-sections of the object to be printed. The vat is lowered every layer, until the whole object is printed. The technique was originally intended as a faster and cheaper way to create prototypes for engineers. In fact, the main advantage of such a technique is the high resolution achievable, since it is based on a laser beam. However, because a specific photopolymer resin is needed, it is costly and does not offer a wide array of materials to print with; even tough new materials are constantly added, and may allow the use of such technique not only for prototyping/ molding, but also for final objects. In the process, supports are needed and must be removed after the process has ended. Cleaning and other post-processing is needed, including curing in UV-ovens, vanishing or blasting with glass beads.

3.2.3.2. Digital light processing (DLP)

This technique is a low-cost version of stereo-lithography. It is based on the same principle of photo polymerization, but instead of a laser beam, it uses a video projector in order to harden the resin. A DLP projector is positioned above a resin vat and the resin is hardened layer by layer, as in the SLA. The results are similar to those of SLA, but here, a higher resolution can be achieved on a smaller projection surface, since the projected image has a fixed resolution (the projector's). The process is cheaper and faster than SLA, since, respectively, it is based on common technology (the beamer) and it hardens each layer at once (Figure 4).

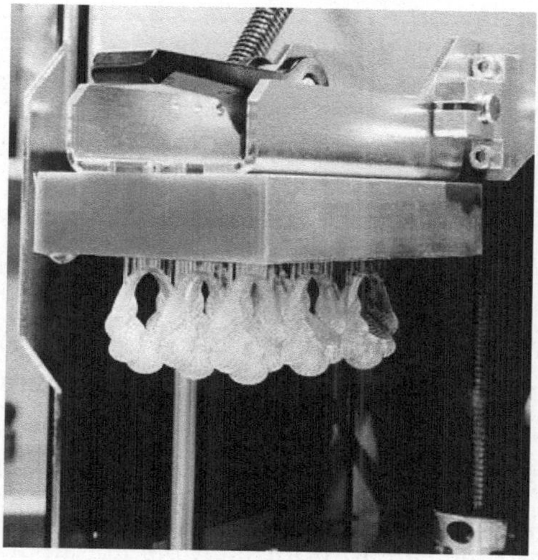

Figure 4. Example of SLA production (https://tinyurl.com/yal2wluv).

3.2.3.3. Multijet modeling (Polyjet)

This technique is a recent development of previous ones. Developed in 2000 by Objet Geometries (now merged with Stratasys), it combines a print-head spraying liquid photopolymers into very thin layers, and a UV lamp—positioned under the print-head nozzles—hardening each of said layers. Layers are created by lowering the work platform, while the head just moves along the Y axis, since it covers the X axis through a number of nozzles. Supports are needed, and printed with a gel-like material by a second row of nozzles. The process also allows the use of a combination between two materials with a varying gradient, thus allowing to locally customize the material properties ('digital materials'). For instance, a mix of soft and hard parts could be printed together. The resolution of this process is also very high. The main drawback is the limitation to photopolymers as printing material, which is expensive and does not offer enough mechanical strength for some uses (Figures 5 and 6).

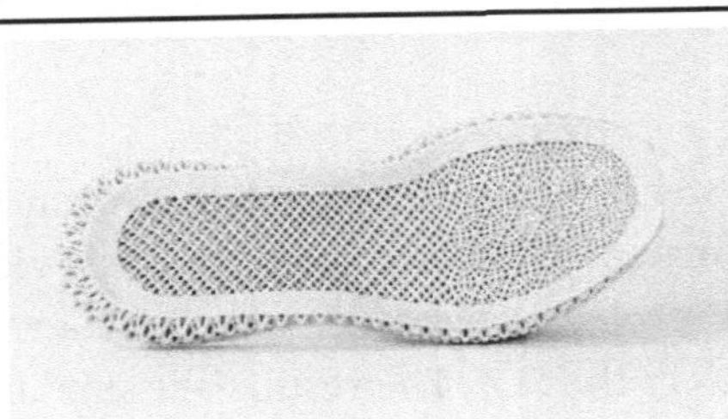

Figure 5. Carbon/Adidas 3D printed sole (https://tinyurl.com/y7ked3cu).

Figure 6. Carbon 3D printed lattice structure (https://tinyurl.com/y8n472bs).

3.2.3.4. Carbon 'digital light synthesis™'

The technique—developed by the 2013-founded company Carbon—uses digital light projection, oxygen-permeable optics, and Carbon's programmable liquid resins and allows printing 'up to 100 times faster than other additive manufacturing processes. [...] Carbon's technology is inherently capable of printing high-resolution parts with an excellent surface finish and isotropic mechanical properties'. It allows 'to print unique lattices that can replace materials such as foam in headsets, shoe midsoles, and seating applications. What is especially unique is Carbon's ability to design and make tunable lattices depending on customer application needs. Engineers for the first time can 3D print multiple unique functional zones within the same monolithic part and tune the mechanical properties within each of these functional zones depending on the application requirements' [5]. This technique is unique in the panorama of additive manufacturing, and it is being used by Adidas to print training shoes' soles on an industrial scale with a variation in the material density throughout, so as to obtain the required local performances. It is then a good example of both the possibilities of 3D printing in the industrial production process of finished goods, as well as the revolutionary potential of obtaining different physical performances by controlling the density and structure of the material.

3.2.3.5. Volumetric 3D printing

'A team of scientists and engineers led by Lawrence Livermore National Laboratory (LLNL) has developed a process that uses hologram-like lasers to make complete objects in seconds inside a tank of liquid resin. Called volumetric 3D printing' [6].

In fact, 'two limitations of additive manufacturing methods that arise from layer-based fabrication are slow speed and geometric constraints (which include poor surface quality). Both limitations are overcome in the work [...], introducing a new volumetric additive fabrication paradigm that produces photopolymer structures with complex nonperiodic three-dimensional geometries on a time scale of seconds. We implement this approach using holographic patterning of light fields, demonstrate the fabrication of a variety of structures, and study the properties of the light patterns and photosensitive resins required for this fabrication approach. The results indicate that low-absorbing resins containing ~0.1% photoinitiator, illuminated at modest powers (~10–100 mW), may be successfully used to build full structures in ~1–10' [7].

3.3. Use in the AEC fields

3D printing in AEC can be seen as an opportunity in many ways.

☆ Direct/indirect (molds).

As we have seen, additive manufacturing is not yet in the stage of fully mature technology, and several new breakthroughs are emerging year after year. This means that there is still a great growth potential, but it also implies that there are still many limitations to overcome, and each currently available technique does not seem to answer many of the needs of industrial production. As to architecture, engineering and construction (AEC), such limitations seem even more problematic. The sheer scale of such endeavors is in fact limiting the kind of techniques that could be adopted to manufacture all or part of a building. Moreover, the requirements for specific physical and mechanical properties—often traditionally obtained through the use of multiple layers of different materials—and the sheer volume of material needed in order to achieve the required performances are other clear limiting factors.

Therefore, depending on the kind of elements to be produced—structural, finishes, etc.—different techniques can be most appropriate. For instance, extrusion of fuse material techniques does not seem appropriate to print huge structural elements, both due to the lack of physical properties and the (low) production throughput.

However, most techniques can be stretched beyond their intended range of production by adopting an indirect approach: for instance, even the said fuse material extrusion techniques can be used to create molds for reinforced concrete structural elements. While scale issues remain, speed and mechanical issues are overcome, since the real structure will consist of the concrete poured in the mold along with steel reinforcements. The main advantage is the opportunity to create, with relative ease, elements that follow complex geometries, which would otherwise be very difficult to achieve, and to do so with great accuracy. Example: ETH mesh-mold, 2014 (Figures 7 and 8) [8].

☆ Modules/components, joints and monoliths.

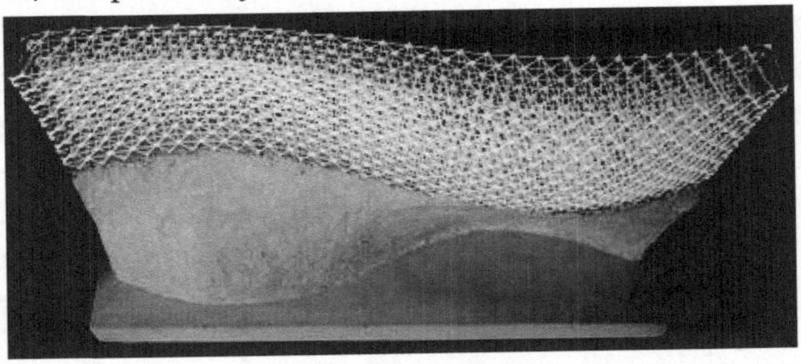

Figure 7. Robotic mesh-molding technique: printed output (https://tinyurl.com/ydddvp4g).

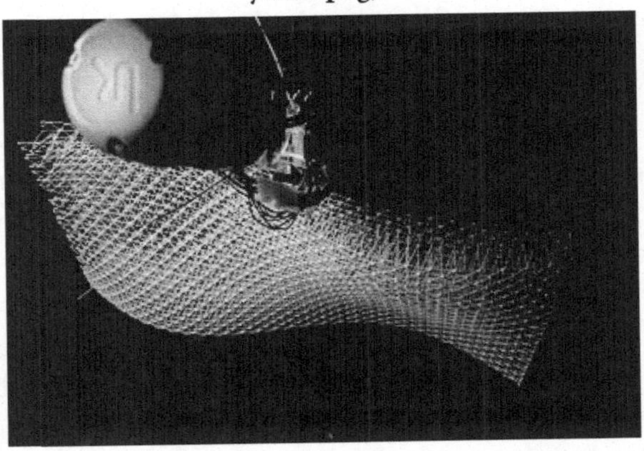

Figure 8. Robotic mesh-molding technique: while printing (https://tinyurl.com/ydddvp4g).

Scale limitations remain one of the main bottlenecks for the use of additive manufacturing techniques in the realm of AEC.

3.3.1. Monoliths

There are some experimental attempts to create and utilize printing machines that could directly deal with the architectural scale, and 'print' entire buildings as 'monoliths', i.e., as a unique piece printed at once, and therefore resulting in an almost seamless unique piece of material(s), possibly with isotropic properties.

An example thereof is 'D-Shape', a company and technique developed in 2004 by the Italian engineer Enrico Dini, where fabrication is possible on an area of 6 by 6 m and limitless height. 'Enrico Dini's printing technique has the great advantage of providing support for overhanging geometries, as sand is selectively transformed to stone within a bed of untouched sand, allowing freeform 3D geometries to be produced. Limitations in this technology are today the strength of materials and printing resolution of approximately 5dpi (20 mm in the X and Y axis and 5 mm in the Z axis)' (Figures 9 and 10) ([4], p. 116).

Figure 9. D-shape printed monolith (https://tinyurl.com/ybdf339t).

Figure 10. D-shape printed monolith (https://tinyurl.com/y8c87o6z).

'Contour crafting' and 'concrete printing' by Loughborough University, as we already saw, are yet other techniques suitable for monolithic production. However, most of the pieces printed so far tend to lack the complexity of freeform 3D geometries, showing a variation along just two of the three axes, thus missing out what is supposed to be one of the main advantages of additive manufacturing. Moreover, concrete printing has been tested with a build volume of just 2 m × 2.5 m × 5 m, which would not fit the required scale for any substantial architectural endeavor.

Besides the specific limitation outlined for each 'monolithic' printing technique, a general criticality lies in the many different kinds of performances required in AEC: mechanical and structural, thermal, permeability to light and air, and the like.

Such aspects are traditionally dealt with by a series of different 'layers' of elements made of different materials. Even a basic bearing brick wall does usually incorporate not only bricks, but also a binder, as well as possibly a damp-proof membrane, etc. Similarly, a reinforced concrete structure usually needs specific layers to deal with the propagation of sounds and vibrations, thermal bridges, and many other aspects. At the moment, it does not seem that additive manufacturing techniques can deal with such requirements effectively, or at least there are clear gaps that must be closed by the extensive use of other techniques. The trade-off between printing resolution and speed is another potential hindrance for this approach to become advantageous: in fact, usually in a building, there is a hierarchy among elements as to their functional relevance, and for some of them it is crucial to be produced with high accuracy and isotropy, while for others speed seems more relevant. Until techniques like concrete printing by Loughborough University—which allows for a change in nozzle resolution while printing, thus controlling the trade-off speed-resolution—will not be industrially feasible and reliable, this approach seems reserved to research projects. However, as we will see, in a future stage of technical development, we can imagine that not only speed and accuracy will be dealt with appropriately for the architectural scale, but also that printing with several materials while gaining control over the fine regulation of the material properties—e.g., density, isotropy and material combination—will make the production of monolithic structure not only advantageous, but even necessary for some advanced new ways of building (Figure 11).

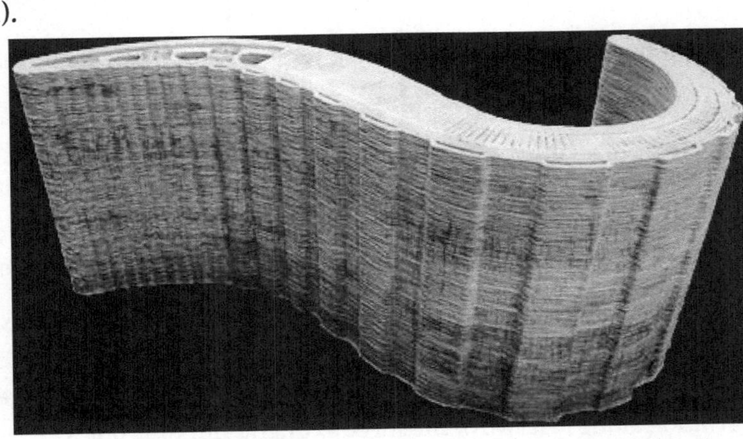

Figure 11. Concrete printing (https://tinyurl.com/ydd8lmsa).

Example of 'monoliths'. MX3D bridge, ongoing. For their nature, bridges and other urban infrastructure may have the right scale for monolithic production. Moreover, they may not require the same number of different performances—notably, the thermal performances required in buildings to guarantee the indoor comfort—and thus may well be constituted by even a sole material. MX3D has chosen a small urban bridge in Amsterdam as an opportunity to showcase and test its 3D printing technology, based on 'multiaxis 3D print technology', a combination of 6-axes robotic arms and metal depositor-welders tipping the robotic arms. 'The robots, which are tipped with welders, will construct the bridge in front of them as they go, literally printing welded steel in midair' [9]. 'The robot arms are similar to those used in the car industry and they can print metals and plastics from single extruders, as well as combinations of the two materials together' [10]. Even though the originally planned on-site printing was dismissed to avoid congestion in a crowded area of the city, the printing method is claimed to be able to create the monolithic structure on site and with no supports/scaffolding, which would open interesting perspectives for the whole AEC field (Figures 12 and 13).

Figure 12. Robotically printed metal bridge (https://tinyurl.com/yalppx2w).

Figure 13. Robotically printed metal bridge (https://tinyurl.com/yanqvhv9).

3.3.2. Modules and joints

Another approach to make the best out of additive manufacturing techniques, especially considering their limitations as of today, is then to use them on a lesser scale, focusing on the specific comparative advantages in creating parts of an architecture.

A first obvious method that has been widely adopted in AEC since antiquity, but especially after the industrialization of the production process, is the decomposition of architectures into modules or components. Such an approach requires that the geometrical subdivision be carefully studied, as discontinuities can constitute weak spots. Moreover, modules need interfaces to be connected to one another, which may imply the need of taking care of multiple layers being connected, while keeping the junction water- and air-tight.

Example: Brian Peters, 2014, 3D printed clay bricks.

'Building Bytes is a project that goes beyond using new tools to make old products. Instead, it follows the additive logic of the printing path—rather than the conventional moulding or extrusion process—to make bricks that are otherwise impractical or impossible to make' (Figure 14) [11].

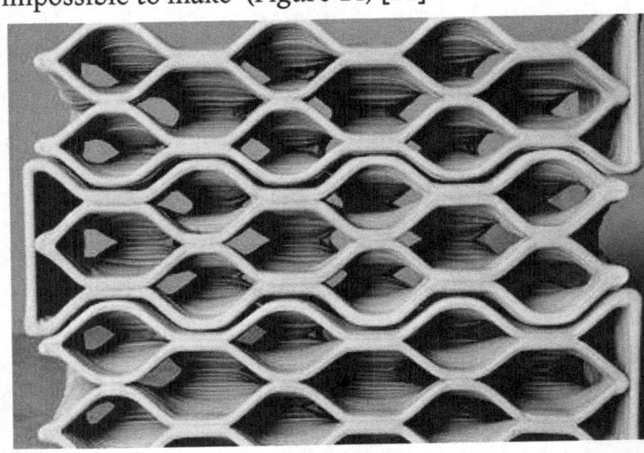

Figure 14. Clay printed modular structure (https://tinyurl.com/yb62v3cz).

A second area where additive manufacturing seems most suitable is the production of joints. Joints are particularly relevant in many contemporary AEC projects since they allow the creation of freeform, irregular geometries by connecting standardized elements. In other words, joints can 'absorb' the geometrical variation of the overall shape by 'internalizing' it in their spatial configuration. The relevance of additive manufacturing techniques then becomes clear, if we consider that it allows producing a number of alike but different elements (mass customization) at the same cost and in the same time than a series of identical ones.

There is more to it: joints typically perform key structural functions, and thus are subject to intense stresses. While this aspect seems to rule out many 3D printing techniques due to the weakness of available materials to print with, the customization of each joint's geometry to meet its specific performance targets can be a crucial success factor of additive manufacturing techniques, allowing optimization techniques such as topology optimization. It is in fact only 3D printing techniques that can give birth to topologically optimized objects with their highly organic and irregular shapes, including voids that would be often impossible to obtain with any other production technique.

Example: Arup, Optimized Structural Element (nodes), 2014–2015 (Figure 15).

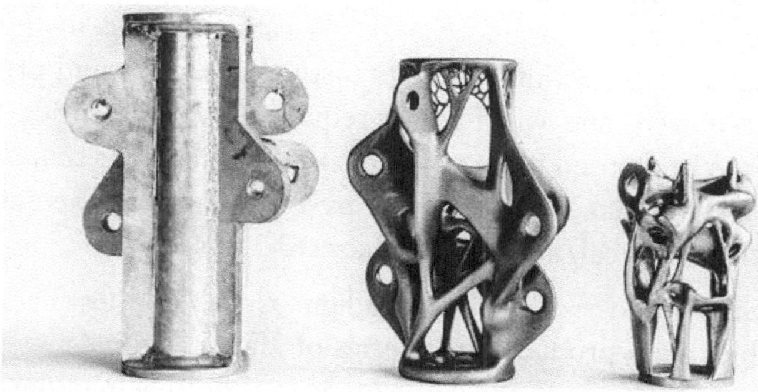

Figure 15. Topologically optimized 3D printed metal nodes (https://tinyurl.com/y9kx2o7l).

The engineering company has successfully produced building structural elements through additive manufacturing, which are an optimization of a standard node for a tensegrity structure. A paper showing the results of the study explains that: 'Based on these initial results the design process was fine-tuned focusing on product integration and improved control of the optimization process. A full set of material tests was executed which should lead to a certification process required for specifying AM-produced products in the Building Industry' [12].

The ability of printing parts that are nested within each other opens up further relevant opportunities for the creation of movable joints. While with traditional techniques, such joints could be created only by welding together parts of the joint—which increases the risk of potential discontinuities and weaknesses—the new approach can produce more uniform pieces of isotropic material (Figure 16).

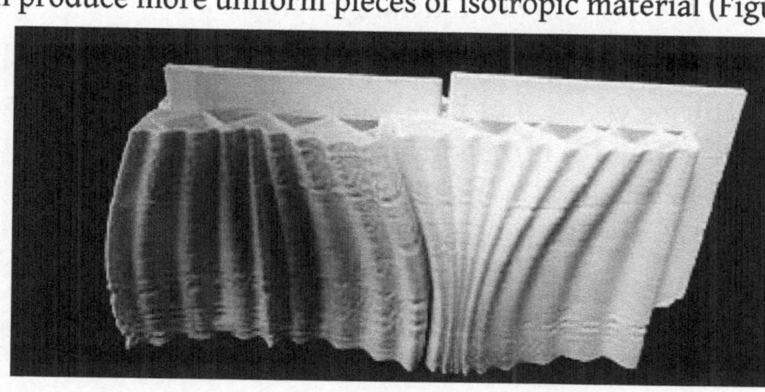

Figure 16. G. Rossi at ACTLab, 3D printed interlocking structure (credits Politecnico di Milano University).

Example: Master Thesis at ACTLab by G.Rossi, 3D printed interlocking structure.

☆ From fast prototyping to 'the real thing'.

The great hype lately surrounding 3D printing and additive manufacturing in general is not only due to the opportunity offered to create complex shapes under the direct control of computers—this has been the case for the last 50+ years in many industrial sectors, such as automotive—but is rather rooted in the now possible production of final construction elements instead of mere rapid prototyping. This evolution has required, and will further require, advancement as to the type of materials that can be printed with, as well as in the tolerances that can be achieved with every new technique, along with the overall quality of the output, including isotropy and other physical/mechanical characteristics.

Working with a rapid prototyping paradigm usually implies that each prototype is tested and then mass-produced in a series of identical copies when the required performances are achieved. On the contrary, using additive manufacturing as a tool for production, it is now possible to create 'final' elements that are different from each other and are produced directly as they are simulated in the software. The tighter correspondence between the virtual modeling and the real output means that—while prototyping remains useful to have some hard data to back the simulations up—most simulations can be now close enough to the real behavior of the printed elements as well as of the overall performance of the structure with no need to prototype each and every element first. Moreover, while using additive manufacturing for prototyping and a different kind of production for the final elements can create a mismatch between the two, the direct use of 3D printing techniques for final production allows to directly harvest the benefits of the technique. The recalled example of Adidas 3D printing shoes' soles is certainly an interesting reference as to the industrial potential of the technique, while the adoption of similar models in ABC at the moment is still mostly in the framework of research.

Example: GE, 2015, fully 3D printed mini jet engine.

This project, which lies entirely within the field of engineering, shows how far additive manufacturing techniques have gone over the last few years. The engine, which could serve a radio-controlled small plane, has been 3D printed using the direct metal laser melting (DMLM) technique. It was then assembled by hand, and is fully working (Figure 17).

Figure 17. GE fully 3D printed working engine prototype (https://tinyurl.com/y85355af).

3.4. Current impact on design/production paradigms and case studies

- ☆ Mass-customization and the ideal match with parametric generative design within the computational design tools.

Additive manufacturing, being a production method transforming the 'virtual' 3D model into a real thing with a comparable cost per volume printed, allows for a wholly new approach to design and fabrication. While the First and Second Industrial Revolutions, as seen, required as much standardization as possible, it is now possible and often preferable to design each element specifically optimized for the function and position it holds within a structure. For instance, façade-shading elements need not be all equal if their position between the sun path and the spaces to shade is different.

In order to exploit this potential fully, however, it would be impractical to design each element one by one: besides being time-consuming, it would possibly be also difficult to calculate and draw by hand what the right geometrical configuration of the said shading elements should be, since the sun path changes over time. Luckily, given the 'direct' creation of any shape from the 3D virtual model, it is enough to devise a system that allows the creation of a series of elements—e.g., our shading element, in the example—that are similar in design and function, but have the right measures for the specific place they are intended for. In other words, we need a tool that—given a set of geometrical and logical relationship between the required performance and the given constraints—could generate a series of optimized elements. Such a tool can be found in the computational design realm, specifically in generative algorithms, where the final shape is generated by the software based on the logical connection and operations between the provided inputs. 'Form is differentiated from the fundamental principles organizing the different elements within the manufactured component. None of the components is considered as an ideal primary model; every element might differ in geometry and form as long as the intricate logical interrelations are accurate. The bigger the variation and the complexity, the higher is the value and the benefit of using an AM machine' ([4], pp. 23-24).

The output of a mass-customization process is then the creation of a series (mass) of industrially crafted objects, which are nonetheless tailored (customized) on the specific place and functions they must perform. If duly performed, such a process can yield specific performances while possibly costing as much as standard elements and using only the material needed.

Example: Politecnico di Milano, Expo 2020 Desert tectonics hypothesis. 3D printed external shading structure in HTPLA + Sand (Figure 18).

☆ Nested and interlocked geometries otherwise impossible to be produced.

Figure 18. Politecnico di Milano, expo 2020 desert tectonics hypothesis (credits Politecnico di Milano University).

Besides movable joints, many other interlocking geometries can be now produced, which would have been at least very complex to craft without AM techniques. Chains, textile-like structures and the like are all examples of interlocking geometries that are usually obtained by knitting thread-like materials, or else require heavy hand crafting, as in the traditional chainmail.

Example: Gürcüm, 2017, textile-like structures.

This study synthetizes the possibilities offered by AM in this area, discussing 'the important properties of traditional fabrics that are to be expected of 3D printed structures namely physical properties like flexibility, bending and and drapability' (Figure 19) [13].

☆ Design from simple 'shape-drawing' to simulation based on material physics and static embedded fabrication constraints.

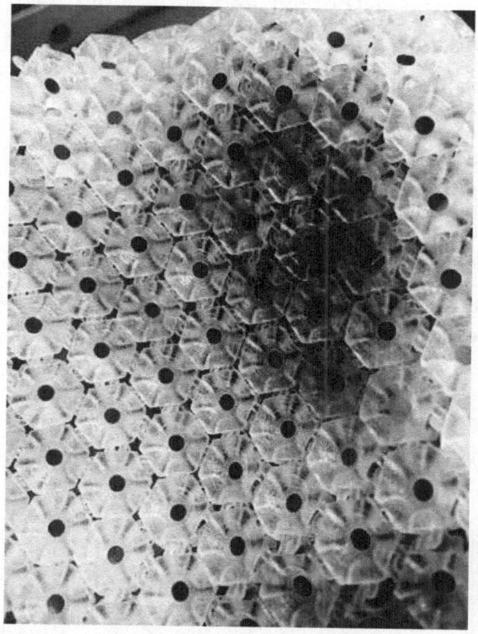

Figure 19. Example of textile-like 3D printed interlocking structure (credits Politecnico di Milano University).

Traditionally, drawing techniques have been used by architects to communicate their project to a series of other professional figures, such as engineers and site managers, in order to have it checked and realized. The shape of buildings and elements thereof was usually devised by the architect in the early design phases based on a rough understanding of the underlying physical characteristics and required performances, and would be further adapted in case the design proved to be impractical to realize. Things have now changed, since it is now possible to include within the design phase a simulation of the physical behavior of the specific shapes based on the specific material characteristics. Therefore, computational design techniques do not just 'represent' a pre-conceived idea of shape, but can allow to reach a shape as the result of a process that incorporates many performance and material constraints, including fabrication constraints.

Example: L. De Sanctis, 3D printed clay brick. The underlying idea of this project was 'to develop a customized design of a very traditional building component: a clay brick. The concept relies on the possibility to have a flexible system of tile modules, which could be site specific and ad hoc buildable with Additive Manufacturing (AM) in relation to the context. The design of the component is developed analytically with respect to a framework of requirements and performance typical of a clay component, with the addition of standard features of a wall system [...] An algorithm developed with the use of Grasshopper and Python has been applied to determine the wall thicknesses and amount of material distributed, while optimizing structural performances of a design and considering production constraints. It has thus been identified as an ideal format (similar to what exists in trade), and compatible with printing constraints, a dimension of 250 × 250 × 120/125 mm. This dimension could also fit within exiting insulating EPS panels (500 × 1000 × 50 mm), integrated with electrical pipes of 8 mm or junction box of 120 × 100 × 70 mm. Another advantage of this system is the possibility of integration within any kind of form or structure, in relation to its use. Due to the necessity to preserve structural equilibrium within a wall, design of cantilevered parts of the brick has been performed within the mass quantity not superior to 40%' (Figure 20) [14].

☆ Optimized and multiperformative shapes as the result of topological optimization and multifactorial design constraints and analysis. Performative biomimicry.

Figure 20. L. De Sanctis, at ACTLab, 3D printed custom clay brick.

As seen, additive manufacturing allows the designers to potentially craft each and every element of an architecture all different from each other. Printing 'topologically optimized' elements seems one of the most valuable opportunities offered by AM. 'Topological optimization is a mathematical approach that aims at optimizing material layout within a given design space, for a given set of loads and boundary conditions such that the resulting layout meets a prescribed set of performance targets. Topology optimization software systematically analyzes the stresses on these shapes and then removes the most superfluous material from the design. This process is repeated over and over by the software until the target amount of material is reached, and by the end the computer design leaves only a skeletal structure. The advantage of parts made with topology optimization is therefore that the same strength characteristics can be created with less material, and this yields a greater strength to weight ratio, an important property across most industries, from automotive, to aerospace, but also architecture and building construction' ([4], p. 157). 'As a practical example, structural rib elements in an Airbus wing designed with topology optimization saved over 500kg in structural weight, which translates to significant cost savings' [15].

Example: R. Naboni at ACTLab, cellular solid lattice structure in pla.

The project combines and applies the principle of topological optimization—deposing material only where it is needed for structural reasons—to a system of load-responsive interconnected struts made of polymeric material. The result is a custom lattice microstructure defined as functionally graded lattice structure, with spatially varying characteristics. 'Algorithms for topology optimization of freeform shapes are employed to determine the material organization as well as a performative matrix [...] The potential of this system relies on its implicit resistance and reduced use of material, combined with the possibility to adapt to any architectural shape. They are composed by an interconnected network of struts, pin-jointed or rigidly bonded at their connections. At one level, they can be analyzed using classical methods of mechanics, a typical space frames, on the other side, within a certain scale range, lattice can be considered as a material, with its own set of effective properties, allowing direct comparison with homogeneous materials. Mechanical properties of lattice materials are governed, in part, by those of the material from which they are made, but most importantly by the topology and relative density of the cellular structure. This methodology requires the description of custom algorithms to generate lattice structures parametrized on the base of a continuous feedback loop from a Topology Optimization and manage the additive process of materialization' (Figure 21) [14].

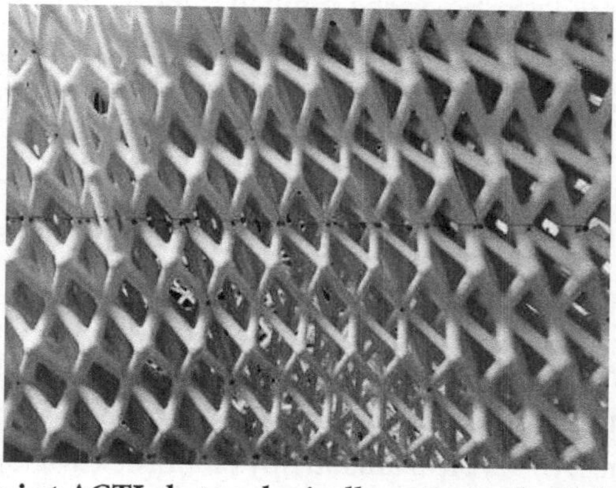

Figure 21. R. Naboni at ACTLab, topologically optimized cellular lattice structure.

Example: Neri Oxman, 2014, Gemini Alpha Chaise Longue.

The inner skin is made of three different rubber-like plastics, printed by a Stratasys' new Objet500 so as to obtain 44 different composites. Each of these composites has a different rigidity and color, and is arranged in a way to cushion the user. The choice of shapes is also informed by their noise-cancelling properties. The chaise is supposed to create a silent and calm environment inside, through the combination of a concave shape reflecting sounds inwards and of the inner surface geometry and materiality, which scatters and absorbs the sound waves (Figure 22).

Figure 22. Neri Oxman, multiperformative 3D printed chaise-longue (https://tinyurl.com/y8f3d6t9).

3.5. Future perspectives

3.5.1. Potential issues

- ☆ Safety certification and accountability (given the 'exotic' nonstandard geometries and the extensive reliance upon software simulation).

One major issue with not only additive manufacturing, but in general any nonstandardized manufacturing, is the implied impossibility to test each and every item produced. This does not constitute a problem in case every set of produced items is exactly the same, since it is enough to test one for testing them all. However, it would not be feasible or at least in sharp contrast to the very same aim of mass-customization—adaptability at a cost comparable to that of standardized solutions—to test all produced items. As already noted, the solution which is usually adopted is to sample-test within the range of produced items, trying to select the right samples along the range. Such process has to rely on the implied correspondence between the virtual and the real, and therefore the specificities of each printing technique have to be accounted for in the software in order to properly simulate the real-world behaviors. For instance, the typical anisotropy of most extrusion-based printing, and the differences in physical/mechanical behaviors along different directions must be factored-in. The process also heavily relies on software-based automatic check about most required performances, e.g., mechanical. The software's results will then depend both on the selection of data that are fed into it, as well as on the reliability of the software itself.

All these issues raise questions about the accountability for any failure of the printed elements, and will probably require a whole new set of legal tools to sort out who is accountable in each case, and where to draw the line between unforeseeable circumstances, due diligence and lack thereof.

- ☆ Optimization vis-à-vis resilience, future adaptation and available architectural langwuage.

Abandoning the paradigm of standardized production for embracing a more and more optimization-driven approach seems a great innovation in view of wasting less material and achieving more with less at the same time. Any structure where the quantity and the structure of the deposed material is optimized to the prospected performances seems like a perfect way to mimic nature and its efficient adaptive behaviors. However—at least until the output of additive manufacturing technique will be re-configurable—3D printed objects, unlike natural living beings, tend to be fixed once and for all, unless additions or modifications are purposively performed. However, such interventions *a posteriori* could clash with the optimized existing (macro- and micro-) geometry, which is based on the specific data and constraints that were decided at the original design and construction time. In fact, such issue holds true for any AEC endeavor where tolerances and redundancies are kept at a safe but minimal level. However, in traditional constructions, the material is usually distributed in a more constant and uniform fashion, so that any slight change to the conditions and constraints might have a less radical impact than on a structure where the material is reduced in quantity and fine-tuned on more specific conditions.

Moreover, also the sheer lack of standard 'interfaces' to join additional elements could be a limiting factor for the adaptation of the structure to new conditions over time. It is quite evident that adding a new row of bricks to a parallelepiped-shaped wall is a much lesser feat than to an Enrico Dini's organic structure.

All this implies that the short-term sustainability of an optimized 3D printed structure could be disadvantageous to the long-term adaptability and resilience of it. A 3D printed structure could be so costly and challenging to adapt, that it might be even cost-effective to tear it down and re-build from scratch. The obvious question is whether such an approach could prove more sustainable at all, and it is wise to imagine that designers and architects will have to include future adaptability within their design thought process, goals and constraints.

Lastly, as regards the architectural language, the use of constraint-based, parametric generative design based on physical properties of materials could raise concerns about the risks of adopting an architectural 'language' as a mere by-product of the chosen design tools. This would also imply that—those tools and techniques being common among designers, and being physical laws constant in time—the formal output of all design processes by different designers and architects would tend to be really similar, if not identical, among each other and over time. This 'end of history', whenever had to become a reality, would of course contradict the essence of millennia of architectural development and the idea of evolution itself. However, it seems reasonable to believe that creativity can and will be shifted to setting the constraints, the performance goals and even the formal relationships that should characterize the space in the view of each architect.

3.5.2. Potentialities

- ☆ **Multiscalar optimization:** from the microscale of materials to the macroscale of architecture.

While, as seen, topological optimization techniques have been already widely used to fabricate a wide range of optimized components and objects in architecture and design, it is still at the level of research that similar optimization methods are adopted as to the microscale of materials. In fact, it is usually the case that topologically optimized shapes are printed with constant material density and composition. 'Most such technologies, however, remain limited to producing single-material, constant-property prototypes from a restricted range of materials' [16]. However, such dualism between the micro- and the macroscales could be surpassed in the near future, since 3D printed structures could be not only topologically optimized at the macroscale, but also at the microscale, both as to density and structure of material, and as to material composition or combination. This would resemble what happens in nature, where the microstructure of plants and other living beings tends to correspond to the required local performances. In fact, 'Since many biological materials are made of fibrous heterogeneous compositions, their multi-functionality is typically achieved by mapping performance requirements to strategies of material structuring and allocation. The shape of matter is therefore directly linked to the influences of force acting upon it. Material is concentrated in regions of high strength and dispersed in areas where stiffness is not required' [16].

Such an opportunity would be a game-changer in AEC. A completely different design paradigm would be needed, since materiality would not be determined *a priori*, as a given property of a commercially available material. Rather, material characteristics—including composition, density and structure—could (and would have to) be engineered as a part of the overall design. Also, the typical discontinuity between micro- and macroscales would have to be reconsidered, since a continuum could be envisaged, similar to a fractal geometry. 'A bio-inspired fabrication approach calls for a shift from shape-centric virtual and physical prototyping to materialcentric fabrication processes. In this approach, not unlike he bones' remodelling process (Figure 7), the distribution of material properties is informed by structural and environmental performance criteria acting upon the component, and contributes to its internal physical makeup. It thus requires a set of virtual and physical prototyping tools and methods that support a variable-fabrication approach, not unlike Nature's' [16].

Voxel-based materials (and printing techniques) would be, in this perspective, another important step forward in such a direction. 'We expect digital materials and the 3D printing thereof to provide unprecedented control over all aspects of bulk materials in diverse fields ranging from micro scale biological tissue constructs to macro scale building projects. The ability to print multiple materials with incompatible processing characteristics and "smart" voxels with specific electrical, mechanical, or fluidic functionality will enable highly functional composite materials to be printed in a simple, robust fabrication process. Over the last few decades, many technologies have benefited enormously by the transition from analog to digital, and we expect the same for three-dimensional matter' ([17], p. 246).

☆ **Multidimensional design:** 4D printing and the inclusion of time and other dimensions to Euclidian geometries.

The shift toward a new production paradigm, where the change in material and material properties is seamlessly possible could open even further opportunities in AEC. In particular, the right combination of materials and material structures can be exploited—according to virtual simulation in the software—to engineer elements, which would behave differently vis-à-vis external conditions, thus determining a change in the overall shape. In other words, the designer could aim for a change of the object shape over time, triggered by some external physical parameters, such as humidity, or temperature change. Similar techniques, which at the moment are still very limited and experimental, have been called 4D printing, with time being the fourth dimension. A pioneering research on the topic has been conducted at MIT's Self-Assembly Lab by Skylar Tibbits, with a specific interest in creating self-assembling shapes [18]. Further research looks at the natural world, especially to the botanical world, in view of creating self-adapting structures: 'Inspired by these botanical systems, we printed composite hydrogel architectures that are encoded with localized, anisotropic swelling behaviour controlled by the alignment of cellulose fibrils along prescribed four-dimensional printing pathways. When combined with a minimal theoretical framework that allows us to solve the inverse problem of designing the alignment patterns for prescribed target shapes, we can programmably fabricate plant-inspired architectures that change shape on immersion in water, yielding complex three-dimensional morphologies' [19].

The use of elements that change over time, adapting to environmental parameters and without any energy requirements or sensors/actuators, could have a profound impact on architecture. For instance, it is easy to imagine a façade, which would adapt to sun and humidity and provide optimal internal comfort without the need of a complex, fragile and expensive system of sensors and actuators.

As an example there of, a prototype of a hygroscopic element that opens and closes reacting to air humidity has been 3D printed and successfully tested by a team of researchers. It could constitute a self-adapting, tunable façade element that does not need neither sensors, nor actuators or energy sources other than the mere change in the environmental conditions. The 'research aims to enhance wood's anisotropic and hygroscopic properties by designing and 3D printing custom wood grain structures to promote tunable self-transformation. [...] A differentiated printing method promotes wood transformation solely through the design of custom-printed wood fibers. Alternatively, a multimaterial printing method allows for greater control and intensified wood transformations through the precise design of multimaterial prints composed of both synthetic wood and polymers. The presented methods, techniques, and material tests demonstrate the first successful results of differentiated printed wood for self-transforming behavior, suggesting a new approach for programmable material and responsive architectures' (Figures 23 and 24) [20].

- ☆ Correspondence virtual-real as implicit requirement for a reliable simulation and appraisal of mass-customized, i.e., nonstandard, elements. Applicability of artificial intelligence as possible solution to the impasse of safety certification of nonstandard elements, as well as performance tuning.

Figure 23. Correa et al., 4D printed adaptive component: when closed (https://tinyurl.com/yd7dy728).

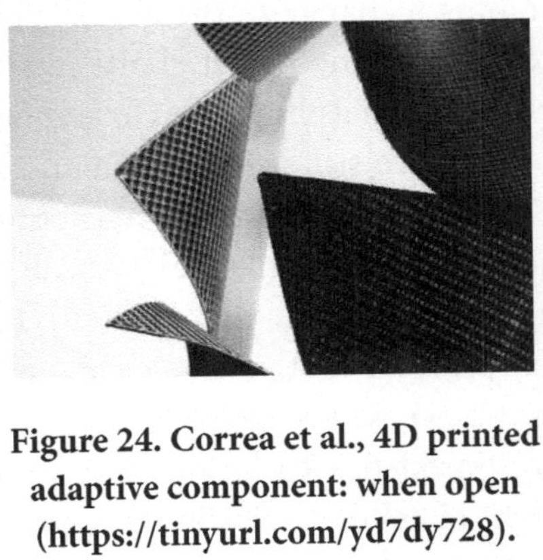

Figure 24. Correa et al., 4D printed adaptive component: when open (https://tinyurl.com/yd7dy728).

The tighter the correspondence between what is being simulated in the virtual realm and what is actually manufactured through 3D printing makes it possible to make assumptions about the performances of the produced pieces. In fact, through sample testing along the range of outcomes, it is possible to validate the virtual simulations about the whole series being manufactured with an acceptable margin of error, possibly with no need to test a sample for each and every shape. Moreover, a feedback loop can help in further perfecting the accuracy of the simulations, attuning them to the specific material and printing machine.

- ☆ The reclamation of the continuous variation in architecture as both biomimicry—hence efficient multiperformative holistic solution—and as an expressive potential.

Allowing architects to design structures which could be continuously changing from the micro- to the macroscale, without having to repeat standardized identical elements, would enable architecture to overcome one of the main criticisms that has been raised since the dawn of industrialization: the loss of those ever-changing spatial qualities which natural refuges—such as caves—and hand-made architecture embedded. Some author ([21], p. 81) noted that 'the "informal" of caves, their anti-geometric character, is transmitted to the "barbaric" culture of the proto-Italian settlement, from Pantalica to Barumini, continuing until our times in the grandiose stone complex of Matera'. Those characters would correspond to those spaces human beings evolved in, and that therefore are programmed to appreciate. Notoriously, the issue of standardization in industrialized architecture was already raised more than a century ago, when authors such as John Ruskin criticized the lack of differentiation among industrially produced architectures, preferring the richness of the nuances provided by historical patina and handwork. '[...] The forms and mode of decoration of all the features were universally alike; not servilely alike, but fraternally; not with the sameness of coins cast from one mould, but with the likeness of the members of one family' [22]. The new production techniques, along with parametric design tools, could mimic nature and hand-making also under this respect, creating a spatial experience more nuanced along a continuum of varying shape and materiality. However, as already noted, this becomes a further challenge for the architects, who need to rethink even their own role and tasks in order to make full and best use of such opportunities.

Example: Digital Grotesque—Hansmeyer, Dillenburger, Zurich, 2013.

This case study does not purport all the advanced materiality and scale-free features we are discussing, but is noteworthy as it shows the degree of continuously varying detailing along a shape. 'Digital Grotesque has been designed through an algorithmic procedure called "mesh-grammars", which procedure consists of rules that articulate the structure out of a primitive input form, by recursively splitting surfaces. The process allows for highly specific local conditions with complex topologies to be generated. [...] The resulting form, consisting of a mesh of 260 million individual facets, has a resolution and level of detail that would be impossible to specify using traditional means, whether drawn by hand or mouse. It provides a glimpse of the potentials of additive manufacturing' (Figure 25) [23].

Figure 25. Hansmeyer Dillenburger, digital grotesque (https://tinyurl.com/yar5udbj).

4. CONCLUSIONS

While additive manufacturing has been around in its main techniques already for some 40+ years now, and cannot be considered an immature technology, it is still undergoing a significant innovation process, often through the hybridization of established base-techniques. Furthermore, in its use—especially within the AEC field—its disruptive potential has yet to be exploited and harvested outside the experimental research or pilot projects.

While waiting for 'the ultimate' technique, some limitations of additive manufacturing can be dealt with through a series of smart strategies. One of these consists in limiting its use to only the parts that need customization, so as to overcome the slower production speed still often associated with AM: very often, it is the nodes that can embed the nonstandard, varying part of the overall geometries, thus allowing for the standardization of all other elements. Another strategy consists in the use of such techniques as an indirect means to support other techniques, as it is the case with 3D printed molds to help create freeform concrete structures. In any case, as it is already the case in some fields like engineering, also within AEC it seems that now additive manufacturing techniques can slowly be adopted not only for rapid prototyping of models and components, or as a mere support technique to other more established techniques, but also to produce functional elements within the final, built structures, or even fully functional entire structures.

The adoption of AM techniques in AEC seems likely to bear some lasting consequences going beyond the 'technical' aspects. In fact, similar to how the invention of press systems changed the role of writers and their professional position, or the recording of sounds created a wholly new environment for musicians, so could additive manufacturing produce a lasting impact on the entire world of AEC, especially as regards the role of the architects and the expressive potentialities opened to them.

On the one hand, it is possible to hypothesize more cost-effective outcomes just by the reduced use of materials and working hours granted by the use of these techniques, as for instance topological optimization, form-finding and other computational design techniques. However, such an approach requires that the architects be aware of the underlying geometrical and physical issues and material properties, both in order to create the final shapes—e.g., as a result of constraint-based design approaches—and in view of 'guaranteeing' safety and durability/ resilience.

On the other hand, it is now possible for architects to regain a wide degree of autonomy (lost with industrialization) as to the creation of nonstandardized elements and custom 'materials', yet within an industrial mass-customized production process. In other words, while the immediate opportunities opened up by AM techniques seem linked just to faster and cheaper production of possibly unconventional and nonstandardized shapes, the greatest opportunities in AEC could lie in the freedom for architects to explore and imagine new languages and solutions that could be at the same time multiperformative in nature, considering multiple constraints and functions, and spatially inspiring.

REFERENCES

1. Kodama H. A scheme for three-dimensional display by automatic fabrication of three-dimensional model. IEICE Transactions on Electronics. 1981;J64-C(4):237-241
2. Kodama H. Automatic method for fabricating a three-dimensional plastic model with photo-hardening polymer. Review of Scientific Instruments. 1981;52:1770. DOI: 10.1063/1.1136492

3. Introducing the world's first 3D-printed car. Local Motors; 2015. Available from: https://tinyurl.com/pk58pkb

4. Breseghello L. A Performative Approach to 3D Printed Architecture, Bachelor Thesis, supervisor Paoletti, I, co-supervisor Naboni, R. Milan, 2015. Available from: https://tinyurl.com/ycpdn68s

5. Carbon web site. [Online] 2017. Available from: https://tinyurl.com/y7ked3cu

6. Thomas J. Volumetric 3D printing builds on need for speed. Lawrence Livermore National Laboratory. [Online] 2017. Available from: https://www.llnl.gov/news/volumetric-3d-printing-builds-need-speed

7. Shusteff M et al. One-step volumetric additive manufacturing of complex polymer structures. Science Advances. 2017;3(12):01

8. Hack N, Lauer WV. Mesh-mould: Robotically fabricated spatial meshes as reinforced concrete formwork. Architectural Design. 2014;84:44-53. DOI: 10.1002/ad.1753

9. Worthington C. Robots to 3-D Print a Bridge From Thin Air. www.wallstreetdaily.com. [Online] 2015. Available from: https://tinyurl.com/ybn7ra5g

10. Robots to 3D Print Steel Bridge in Mid-air|MX3D. www.arch2o.com. [Online] 2018. Available from: https://tinyurl.com/yd2sp6ru

11. Shapiro Fink G. Award: Building Bytes. www.architectmagazine.com. [Online] 2014. Available from: https://tinyurl.com/ycaz6joq

12. Galjaard S et al. Optimizing structural building elements in metal by using additive manufacturing. In: Proceedings of the International Association for Shell and Spatial Structures (IASS). Journal of the International Association for Shell and Spatial Structures. 2015. Available from: https://tinyurl.com/y72l4xmb

13. Gürcüm BH et al. Implementing 3D printed structures as the newest textile form. Journal of Fashion Technology & Textile Engineering. 2018;S4:019. DOI: 10.4172/2329-9568.S4-019

14. Paoletti I. Cultura industriale e progetto contemporaneo: Esempi di sistemi costruttivi sperimentali. Journal of Technology for Architecture and Environment. 2017;13:295-305. Available from: https://tinyurl.com/ybb9qz44

15. Jaffe BH. Topology Optimization in Additive Manufacturing: 3D Printing Conference (Part 5). On3D Printing; 2013. Available from: https://tinyurl.com/yaw5ea57

16. Oxman N. Variable property rapid prototyping. Virtual and Physical Prototyping. 2011;6(1):3-31. DOI: 10.1080/17452759.2011.558588

17. Hiller J, Lipson H. Tunable digital material properties for 3D voxel printers. Rapid Prototyping Journal. 2010;16(4):241-247. DOI: 10.1108/13552541011049252

18. Tibbits S. 4D printing: Multi-material shape change. Architectural Design. 2014;84:116-121. DOI: 10.1002/ad.1710

19. Gladman A et al. Biomimetic 4D printing. Nature Materials. 2016;15:413-418

20. Correa D et al. 3D-printed wood: Programming hygroscopic material transformations. 3D Printing and Additive Manufacturing. 2015;2(3):105-117. DOI: 10.1089/3dp.2015.0022

21. Zevi B. Storia e controstoria dell'architettura in Italia. Rome: Newton Compton; 1997

22. Ruskin J. The Stones of Venice, Volume II, Section XLVI, Chapter VII. Gothic Palaces. London, Smith: Elder & Co; 1853

23. Hansmeyer M, Dillenburger B. Digital Grotesque Towards a Micro-Tectonic Architecture, in Serbian Architectural Journal (SAJ). Architectural Education in the Post-Digital Age. John Wiley & Sons Ltd. Images; 2013;5(2):194-201

CHAPTER-4

PRODUCTION MANAGEMENT FUNDAMENTALS FOR ADDITIVE MANUFACTURING

1. INTRODUCTION

The industry is focusing its attention on additive manufacturing (AM) because it seems to be the way to realize a new industrial revolution. The first contribution about the AM appeared during the 90s [1] with a work by Massachusetts Institute of Technology (MIT); in this chapter, a first attempt to define the problem of the 3D manufacturing and its integration with the traditional way to produce the objects was described.

After this first contribution and many years, the 3D printing, or as it is named in a more detailed way the AM, arrived to the practitioners' world and many of the theoretic hypothesis and problems became real. The application of this technology in real cases or in the technological laboratories let to burst not only a lot of opportunities but also a lot of criticalities in its application.

In last years, the AM is employed in some first pilot production systems for the aerospace and aeronautic sector with application laboratories in collaboration with the MIT [2, 3]. Obviously, their productions are focused on small volume products since it is just a first attempt of experiments in this new field. From these first applications the industries, together with primary universities, are trying to understand the applicability of this new technology to substitute or at least to integrate their traditional production systems with these new ones, recognizing in the AM an opportunity to optimize their processes with particular attention to the design, engineering, production, and logistic.

The main problem for the practitioners and researchers, nowadays, is how to integrate this technology with the old, and which are the new paradigm to optimize the production using it.

The aim of this chapter is to present firstly a literature review that is able to cover three relevant sectors of the industrial systems management problem, two of these are related to the possibility to apply this technology to the industrial world (analyzing the mechanical characteristics of the materials worked and the tolerance of working achievable through AM) and a third, that is more referred to the management methods for the AM and in particular the measurement of costs for the processes of AM production methods and the scheduling models if present. After this literature review, a possible model to measure the costs using the AM will be presented and after a mathematical model to schedule the AM activities will be presented, giving the possibility to the reader to understand the decision problem and so to apply the resolution method preferred by him.

2. LITERATURE REVIEW

Firstly, we will start from the materials and their resistance using AM. This is the first topic investigated in the international literature since its investigation is the first point to demonstrate the feasibility of this kind of technology in the industrial world.

The keywords investigated were several and reported below: (i) AM materials, (ii) AM tolerances in production, (iii) AM metallic materials, and (iv) AM polymeric materials and similar or combinations of the previous words.

The materials more studied using AM are the titanium alloys, even if also other materials are studied such as the Cr-Co alloys, stainless steel, and other metallic materials.

The first investigation about the materials is about the tolerances in production. This is a very felt issue in the production environment since it demonstrates the feasibility of the process, from a first point of view, and the need to integrate the technology with other old technologies to realize a finished part.

The part orientation was demonstrated as influencing variable in the tolerances realization using AM since 2011 as it was reported in a paper in that year [4]. The paper focused the attention on the geometric tolerances related to the orientation of the part and, in fact, it was the first to do so. Before this study other study appeared, for example, the one by (i) Arni and Gupta in 1999 in which the planarity tolerance using an AM technology [5] was investigated and by Hanumaiah and Ravi for other linear geometric tolerances [6]. The tolerances were investigated also in relation with the production parameter by another study [7] that investigated the circularity in the AM in relation with the cutting angle of the starting point (i.e., the error is minimized if the starting cut angle is equal to 0°); going in deeper analysis on this relation, it is also possible to find another paper about the relation between geometric tolerances and production parameters by Lynn-Charney and Rosen [8] who studied and defined a new decision support system (DSS) capable to minimize the errors for positioning of the part in the production chamber and for some geometric tolerances.

Another problem in the AM production is the surface definition in the software file of the part, a paper that investigates this issue appeared in 2016 [9], in this chapter, it is defined a mathematical model to minimize the dimensional errors. Using this model, in an experimental campaign, the reduction of the 70% of the surface defects was eliminated.

Other authors investigated the effects on geometrical tolerances due to the thermal deformation of the melting process for the materials in the production chamber [10]. Other researchers [11] investigated the geometric errors of cylindrical shape dependent by the passage from the CAD model to the triangular shape for stereolithography and they built a new procedure to minimize it.

The geometrical and dimensional tolerances are a topic very felt also in other sectors such as the medical products. In this sector, it was found a contribution of 2015; in this chapter, it performed a sensitivity analysis on the several possible causes of the dimensional and geometrical errors that are possible in the production of this particular kind of objects [12]. The main elements of problem are (i) the quality of the image acquisition and printing, (ii) triangulation density, and (iii) segmentation threshold.

From the papers presented before, it is quite easy to understand that the AM technology is capable and maturely produces parts with good quality, so it is applicable to the industrial sector.

Once that the chapters on the tolerances in production were analyzed, they will be analyzed chapters referred to the mechanical properties of the materials produced with the AM technology.

The Ti-6Al-4 V material is used in many aerospace and mechanical sectors. In fact, as it is reported in a publication of 2015, the alloy Ti-6Al-4 V AM production is good to improve the *buytofly* index from a typical value of 15 to 1 for the aerospace industry [13]. This ratio is obtained using the raw materials compared with the weight of the components at the end of their production. For the Ti-6Al-4 V, using an electron beam melting (EBM) method is possible to have a deposition rate of 500 mm/s, with a moderate operational cost. In the paper by Szost, they reported the following defects for the EBM:

- ☆ Vacancies of melting
- ☆ High-porosity of the materials.

The porosity of the materials produced with AM is a very important issue and this is evident since 2011 when a paper by Baufeld and others appeared [14]. In this chapter, it is fixed a limit value for the porosity of the materials to avoid the gas capture phenomenon and this limit is 6%.

Another paper reports the results of several mechanical tests of the alloy Ti-6Al-4 V when produced with AM (laser-beam deposition (LBD) and the shaped metal deposition (SMD)). In particular, it is interesting that for both the products it is not present any particular remark about the comparison with the traditional ones; they have a critical break tension of 900–1000 MPa and a starting tension for the plastic behavior to 770 MPa.

The production methods such as the EBM, the LBD, and the SMD characteristics are summarized in Table 1.

Table 1. Production parameters used for EBM, LBD and SMD in Baufeld et al.

	EBM	LBD	SMD
Max power (kW)	3	3.5	2.2
Max welding velocity (mm/s)	16.0	10.0	5.0
Max wire-feed velocity (mm/s)	25	40	33
Max wire diameter (mm)	1.0	1.2	1.2
Max height deposition step (mm)	0.1	1	1
Max deposition rate (kg/h)	0.1	0.7	0.6
Max wall thickness (mm)	10	4-5	9.1

To remark what before stated it is possible also to refer to a paper appeared on 2015 [15] on the mechanical properties of the alloy Ni-Ti produced using the technology of shaped memory alloy (SMA). In this chapter, the controls performed are several reported below:

- ☆ Scanning electron microscopy (SEM)
- ☆ Differential scanning colorimeter (DSC)
- ☆ X-ray diffraction (XRD)
- ☆ Micro-hardness test.

In all the tests, the products produced with the Ni-Ti alloy with SMA presented a very close behavior to the ones produced with traditional methods, confirming the goodness of the AM methods.

In 2015, a chapter focused its attention on the laser powder deposition (LPD) method that is very used in the spare parts prototyping; the paper mainly investigated a method to recognize defects during the production run that is ongoing [16].

Always on 2015, another paper appeared demonstrated the goodness of the parts realized in titanium alloys using the AM [17].

Another material used with the AM technology is the AlSi10Mg, that is, a no-metallic material. This material, when produced using the AM, has very similar mechanical characteristics when compared to the traditional production methods, as it was demonstrated in a paper where this fact was demonstrated using a DMLS technology [18].

From the literature analysis about the mechanical characteristics of the materials worked with AM technology is quite evident that the materials worked with this kind of technology are comparable at all with the traditional ones, except for some defects about the porosity of the materials, that can be faced with several adjustments and practices in the cycle or using a post-processing phase with traditional technologies.

Therefore, in conclusion, speaking in terms of quality of the final product, the AM processes are able to produce good parts, and so, it is possible to consider it in a production environment.

After the presentation of the process capability to realize good parts, let us continue with the operations management literature review.

Actually, different attention has to be paid to the operations management themes, since it is a research field for the AM not so much developed as the one about the materials and tolerances.

The operations management are the themes related to the: (i) production organization, (ii) production balancing, (iii) production quality, (iv) life cycle management, and (v) production sustainability.

The keywords investigated were strictly connected to the themes above mentioned with the adding of the AM words.

For the purpose of this chapter, they have analyzed 16 papers from several journals of the main scientific data base available nowadays.

A first contribution analyzed is from an Indian study about the costing in the operations for the prototypes production with metallic materials. In the paper [19], it is evidenced that the rapid prototyping (RP) finds great advantages from the AM application since it allows reducing dramatically the contribution of the production costs for a single piece.

Another contribution also in 2010 appeared in USA, the paper investigated the sustainability impact of AM production system [20]. In the paper, in particular, it is presented an assessment framework for the sustainability assessment for an AM production system, considering the following elements: (i) energy consumption, (ii) waste production, (iii) water usage, and (iv) environmental impact.

After these two first papers, for some years no particular research paper was published, maybe because the theme was too unripe. The theme was looked with interest once again in 2013, that is, after few years with the rising of the first possibility to bring the AM in the industrial context.

Actually, in 2013, they appeared some papers of interest for the AM operations research theme. In this year, they appeared a paper about a cost-benefit analysis between traditional and AM methods for RP operations performed [21]. The benefits are dominant to the costs when the AM is used instead of traditional production systems.

In 2014, another study appeared to prove the convenience of an AM, from the supply chain management point of view [22]. It is defined a decision framework for the application of this technology to a specific product, and the AM resulted in the best choice, especially when the product to be produced is very complex in shape.

Always on 2014 another implementation framework for the AM technology was proposed by another research group [23]. In this paper the authors tried to understand if there is a model of decision capable to help the technology selection using also the AM in the production environment. Another contribution about a very similar theme appeared in 2015 by the US-Chinese research group demonstrated the convenience of the AM considering the supply chain and life-cycle cost models [24].

The technology selection decision was also analyzed by another paper published on 2014 [25]. The framework is based on the evaluation of parameters such as the part complexity, the quantity to be produced and the level of customization that the product requires.

With the 2015 the research started to be more focused on the operations management optimization. Indeed, on 2015 the creation of DSS able to put at the right level of a bill of materials a part produced using the AM technology appeared; the model does this minimizing the life cycle cost and maximizing the value chain of the product [26].

On 2015 another paper focused on the insertion of the AM technology in traditional production systems appeared [27]. The paper deals with the identification of the main facets to be solved to make easy the AM insertion, referring principally to the operations management issues.

Always in the same year, a paper about the economic consequences of the AM use in a production company appeared [28]. The paper depicted the main economic variables applicable to the AM, such as the marginal costs of the production, the quality impacts on production economics, and so on. Very similar works about the economic facets related to the AM implementation in traditional production systems are present also in other paper such as [29, 30].

Another great theme about the economics issues related to the AM was investigated in another paper appeared on 2015. This paper faces the problem to define a selling price for an AM built part; this was done using the gray theory [31].

Also, the environmental and the other variables of the sustainability were deeply investigated as demonstrated from several publications in the international literature [32, 33].

Another theme investigated in the international literature is about the possibility and the framework to make feasible the integration of an AM technology with the traditional subtractive machines [34].

Therefore, the AM operations management facets are studied since the 2010 but they started to be mature in terms of study only from 2013, when some first publications about the possibility to analyze the AM from the costs and operations management perspective started to appear. After this year, (i.e. 2013) the number of papers in this field exponentially started to grow up (see Figure 1).

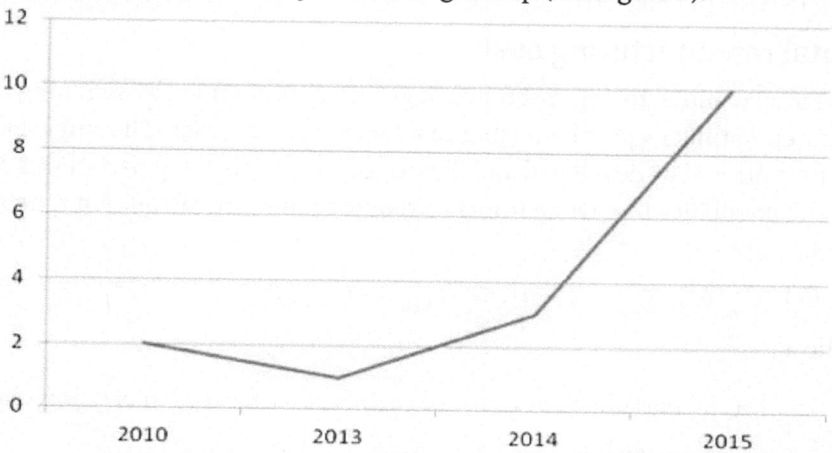

Figure 1. Number of operations management per year.

Nevertheless, it is quite obvious from the literature review analyzed that the operations management using AM is not well covered yet from the research point of view. Actually, this lack is well understandable from Figure 2, in which the operations management theme is reported in terms of number of papers appeared in the years.

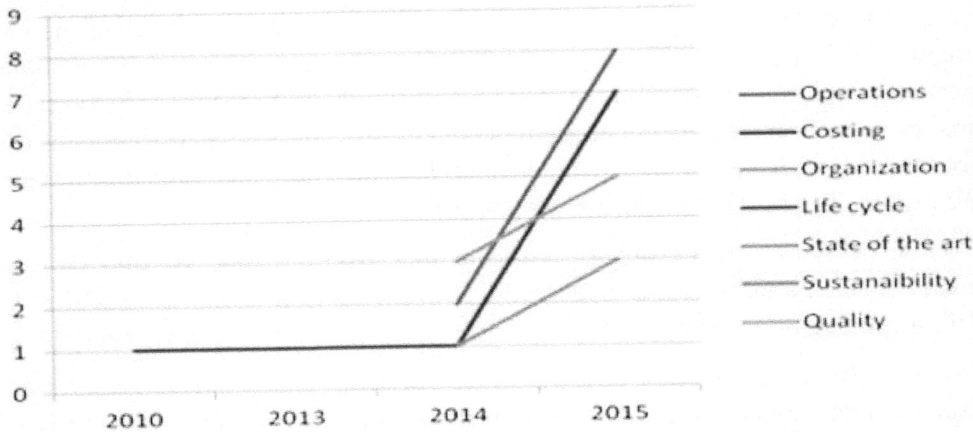

Figure 2. Number of publications per year and per theme.

The first points to be developed in the future are mainly focused on the definition of a proper cost accounting model and also a scheduling model for the AM.

The aim of this chapter is to introduce the reader to some first definition attempts by the author about these two specific issues.

3. THE PRODUCTION COST MODEL

The present model is summarized and reported by another paper of the authors appeared on 2017 [35, 36, 37].

3.1. Total manufacturing cost

The total manufacturing cost for each geometry is obtained by summing the cost of each step to finish a part belonging to a specific client order. The single elements of this formula will be developed and illustrated in the following part of this section, giving the possibility to anyone wants to replicate the cost calculation scheme here reported.

$$C_{tot}(G_i) = C_{prep}(G_i) + C_{buildjob}\,G_i + C_{setup}\,G_i + C_{build}\,G_i + C_{removal}\,G_i \qquad E1$$

Where:

C_{tot} : ToC tal manufacturing cost of each part with *i*th geometry [€/part].

G_i : *i*th geometry [-].

C_{prep} : Cost for preparing geometry data (orientation, support structures, etc.) [€/part].

$C_{buildjob}$: Cost for build job assembly [€/part].

C_{setup} : Machine setup costs [€/part].

C_{build} : Cost for building up a part with *i*th geometry [€/part].

$C_{removal}$: Cost for removing the part with *i*th geometry from the machine chamber [€/part].

3.2. Cost for preparing geometry data

The first step of the process reported here is the design of the geometry data that includes orientation and support structure generation for each geometry. A possible formulation for this cost considers the specific value of a cost referred to the number of parts to be produced for each geometry, as reported in (2):

$$C_{prep}(G_i) = \left(C_{op.pre} + C_{PC}\right) * \frac{T_{prep}(G_i)}{N_i}$$

Where:

C_{prep}: Cost for preparing geometry data (orientation, support structures, etc.) [€/part].

G_i: *i*th geometry [-].

$C_{op.\,pre}$: Pre-processing operator's hourly rate [€/h].

C_{PC}: Hourly rate of the workstation including costs of required software and tools [€/h].

T_{prep}: Time required for preparing CAD data [h].

N_i: Quantity of the part with *i*th geometry [-].

3.3. Cost for building job assembly

In the traditional approach, as reported in Rickenbacher et al. [38] this cost is allocated equally between all parts while in the present formulation this cost is allocated with a ratio that considers the how much volume each geometry occupies in the total volume given by the geometries inserted in the build camera, as reported in (3):

$$C_{buildjob}(G_i) = T_{buildjob} * \left(C_{op.pre} + C_{PC}\right) * \frac{V(G_i)}{\sum_i V(G_i) * N_i}$$

Where:

$C_{buildjob}$: Cost for build job assembly [€/part].

G_i : *i*-th geometry [-].

$T_{buildjob}$: Time required for build job assembly [h].

$C_{op.\ pre}$: Pre-processing operator's hourly rate [€/h].

C_{PC} : Hourly rate of the workstation including costs of required software and tools [€/h].

V : Volume of the geometry [cm³].

N_i : Quantity of part with *i*th geometry [-].

3.4. Machine setup costs

When all the previous activities, that is, the preparing geometry and the planning phase are completed the real production phase can start. This phase includes the data import and machine setup phases. During this time, the machine cannot be used, and for this reason, we included its hourly cost. Also, in this case, we used the parts volume like the allocation criteria and the final formulation is as the one reported in (4):

$$C_{setup}(G_i) = \left(C_{op.mach} + C_{mach}\right) * \left(T_{setup} + \left(F_{mat.ch} * T_{mat.ch}\right)\right) * F_{inertgas} * \frac{V(G_i)}{\sum_i V(G_i) * N_i}$$

Where:

C_{setup} : Machine setup costs [€/part].

G_i : *i*-th geometry [-].

$C_{op.\ mach}$: Machine operator's hourly rate [€/h].

C_{mach} : Machine cost per hour [€/h].

T_{setup} : Time required for machine setup [h].

$F_{mat.\ ch}$: Factor to model the frequency of material changes [-].

$T_{mat.ch}$: Time required to change material [h].

$F_{inertgas}$: Factor to model extra effort required for handling in protective gas environment [-].

V : Volume of the geometry [cm³].

N_i : Quantity of part with *i*-th geometry [-].

In the previous formula, it is also possible to include the possibility to work with an extra time of processing due to the use of protective gas ($F_{inertgas}$). Its value can either be 1 or 0. Also, the change of material can be considered using a 0/1 variable named ($F_{mat.ch}$). Furthermore, if the costs have to be divided on more build jobs, a fraction can be used in the formulation.

Machine cost per hour is obtained by dividing the machine purchase cost by the machine depreciation period and its uptime per year:

$$C_{machine} = \frac{Machine\ cost}{h * upt}$$

Where:

$C_{machine}$: Machine cost per hour [€/h].

Machine cost : Machine purchase cost [€].

h : Machine depreciation period [years].

upt : Machine uptime [hours/year].

3.5. Cost for building up a part

After the presentation of the previous parts of the total production cost, let us to introduce the formula for the calculation of the building step. In this phase, the machine concurrently builds all of the parts in the chamber. The cost's items are:

- ☆ Machine
- ☆ Energy
- ☆ Material
- ☆ Gas

Therefore, it is possible to define the (6), that is the building cost formulation, which also includes a waste factor for the powder used in the deposition and sintering phase:

$$C_{build}(G_i) = T_{build}(G_i) * (C_{mach} + C_{inertgas} * Gas_{cons} + C_{energy} * P_{cons} * K_u) + M(G_i) * (C_{material} * W_f)$$

Where:

C_{build} : Cost for building up a part with *i*th geometry [€/part].

G_i : *i*-th geometry [-].

T_{build} : Total building time [h].

C_{mach} : Machine cost per hour [€/h].

$C_{inertgas}$: Cost of inert gas [€/m³].

Gas_{cons} : Average gas consumption [m³/h].

C_{energy} : Mean energy cost [€/kWh].

P_{cons} : Power consumption [kW].

K_u : Utilization factor [-].

M : Mass of the geometry [kg].

$C_{material}$: Material costs [€/kg].

W_f: Waste factor for powder [-].

3.6. Cost for removing a part from the machine

When the operations of building up are concluded, it is necessary to remove the objects and the substrate plate from the machine chamber. Also, in this case, we included a factor to model, that is, capable to consider the extra time effort for handling in a protective gas environment the production phase. The allocation criteria of this cost are based on parts volume. The formula for this addendum of (1) is reported in (7):

$$C_{removal}(G_i) = T_{removal} * \left(C_{op.mach} + C_{mach}\right) * \frac{V(G_i)}{\sum_i V(G_i) * N_i} * F_{inertgas}$$

Where:

$C_{removal}$: Cost for removing the part with *i*th geometry from the machine chamber [€/part].

G_i : *i*th geometry [-].

$T_{removal}$: Time required for removing parts from the machine chamber [h].

$C_{op.mach}$: Machine operator's hourly rate [€/h].

C_{mach} : Machine cost per hour [€/h].

V : Volume of the geometry [cm^3].

N_i : Quantity of part with *i*th geometry [-].

$F_{inertgas}$: Factor to model extra effort required for handling in protective gas environment [-].

Substituting the single elements of the (1) with the equations from (2)–(7), it will be possible to consider the total cost formulation.

4. THE SCHEDULING PROBLEM FOR ADDITIVE MANUFACTURING (AM)

The AM scheduling is a problem, different from the traditional single machine scheduling, to be solved since this technology started to be a permanent part of the production environment of several companies especially in the field of defense and aerospace. The model presented here is a summary of what was presented in [39, 40]. The question to which the paper wants to answer is always the same of all scheduling problems, that is:

"What's the schedule that allows to respect due dates with the least production cost?"

The question is the same but the context as explained in the introduction is very different from the traditional one for the motivations are related to the set-up, that is no more present in the traditional form since it is done all through the design phase and the transferring of the data from the design workstation to the machine and since the fact that with AM it is possible to produce several kinds of geometries in the same production run. For this reason, let us introduce a multi-objective model for the AM schedule that is able to consider also the new constraints given by the new context.

The development framework (Figure 3) is like the ones for the traditional problems of *production scheduling, so the* production orders are the inputs of AM machine scheduling problem and each order is characterized by the following attributes:

d_i:	demand of G_i - th geometry or PN	part
dd_i:	due date of G_i - th geometry or PN	day
V_i:	volume of G_i - th geometry or PN	cm^3

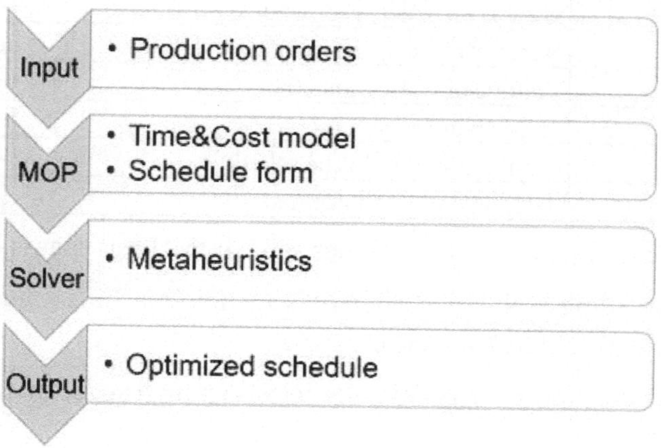

Figure 3. Mathematical model frame.

After that the attributes for the production orders are listed, it is worth to note that in this paper a time and cost model will be applied, in particular, they will be considered the Completion Time (CT) and the Total Part Cost (TPC). CT is the time to produce a single unit of G_i - th geometry, while TPC is the costs to be covered to produce a single part.

Once the main description elements of our model is described let us to introduce the mathematical formulation of the optimization problem here analyzed. The basic model is taken from a research paper, that used earliness and tardiness as objective function [41], to these objectives in this proposal it is added the cost:

$$F_S = F_{ET} + F_{CP} \qquad \textbf{Min!}$$

$$F_{ET} = \sum_{i=1}^{n_g} [\alpha_S E_i + \beta_S T_i]$$

$$F_{CP} = \sum_{i=1}^{n_g} \gamma_S TOC_i$$

Subject to:

$$\sum_{i=1}^{ng} n_{i,j} * V_i \leq V_{chamber} \quad \forall j \in [1, n_b]$$

$$\sum_{j=1}^{nb} n_{i,j} = d_i, \quad \forall i \in [1, n_g]$$

$$\alpha_S, \beta_S, \gamma_S, TOC_i, V_i, V_{available} \in \mathbb{R}^+$$

$$E_i, T_i, i, j, n_g, n_b \qquad \in \mathbb{Z}^+$$

Where:

α_S :	Earliness constant weight	1 / day
β_S :	Tardiness constant weight	1 / day
E_i :	Earliness of i – th geometry	day
T_i :	Tardiness of i – th geometry	day
TOC_i:	Total Order Cost of the *i-th* geometry	[€]
γ_S :	Cost constant weight	[1/€]
n_g :	number of order/geometries	-
$n_{i,j}$	Number of the i – th item in j – th build	part
V_i	Volume of i – th geometry	cm^3
$V_{chamber}$	Build chamber volume	cm^3
n_b	Number of build in the schedule	-
d_i :	demand of G i – th geometry	part

The proposed scheduling model has some hypothesis that is listed below:

☆ The scheduling problem faced here is a Single-machine scheduling problem, where the machine is an AM machine.

☆ The part orientation is given and there is enough space to manually remove the part.

☆ The build chamber allows construction of parts on top of each other by support structures or other solutions.

☆ Stock costs are neglected.

To solve this kind of problems that are generally reported as *NP-hard* problems, it is possible to apply several kinds of heuristics such as Tabu Search, genetic algorithm, simulated annealing, ant colonies, bees, particle swarm optimization, and so on.

5. CONCLUSIONS

In this chapter, an attempt to analyze and categorize such themes related to the production issues of the AM technology was done. AM technology began becoming an industrial solution recently, and so it was recognized as an interesting theme with a possibility to deepen the research area if this technology achieved a good level of maturity in terms of mechanical resistance characteristics and tolerances achievable. After the first step analysis, the state of the researches about the operations management field was analyzed and investigated.

This chapter defines a precise way for the literature review, and it analyzes and categorizes several papers with the method introduced before.

From the literature analysis conducted, it was quite clear that from a technological and from a production quality level (measured from the capability to achieve such levels of geometrical and dimensional tolerances) not so many issues are open and so it is possible to conclude that the AM is quite ready to be brought in the industrial context, given some adjustment still needed about the processes to avoid the porosity problems for the parts worked with the AM technology and to achieve some geometrical tolerance as the concentricity and circularity. Therefore, the pieces in a metal can be realized directly using the AM technology, having a good level of reliability of the part, when it is under mechanical stress. Moreover, the parts realized in this way are good in terms of tolerances achievable, limiting the need of mechanical post-processing activities.

Starting from this result in this paper it was investigated if a scientific literature lacks on the possibilities of the AM to be a production technology in the industrial world exists or not. In fact, to make real the possibility to effectively bring this technology in the production context, it is required that the AM becomes more and more a theme studied and known from the operations management point of view, also offering the possibility to be measured in a production context. In fact, the technology is evolving, and the need to understand the production management issue is becoming a need. Therefore, it is possible to say that if from the point of view of the mechanical tolerances and properties a lot of work was performed and good results are available not the same can be said from the management point of view. Therefore, this state of the art can individuate as the first point of study an important issue about the methods to measure the process costs when an AM technology is used to produce an item.

Moreover, as argued in the previous paragraph, the management theme seems to be affected from a very important absence. Many authors have begun to study the management issues related to the general systems; however, nobody recognizes a main limit in the actual knowledge level.

Therefore, in this chapter, it was presented a cost allocation model that fits the requirements of this new technology and also a mathematical formulation for the scheduling problem of a single AM machine is presented.

REFERENCES

1. Cima M et al. Three-dimensional printing: The physics and implications of additive manufacturing. Annals CIRP. 1993;42(1):257-260
2. LaMonica M. 2013. https://www.technologyreview.com/s/513716/additive-manufacturing/ [Online]
3. Duerden D. 2011. http://www.3d-printing-additive-manufacturing.com/media/downloads/54-d1_1220_b-david-duerden_mbda.pdf. [Online]
4. Ratnadeep P, Sam A. Optimal part orientation in rapid manufacturing process for achieving geometric tolerances. Journal of Manufacturing Systems. 2011;30:214-222
5. Arni R, Gupta SK. Manufacturability analysis of flatness tolerances in solid freeform fabrication. Journal of Mechanical Design. 1999;123(1):148-156
6. Hanumaiah N, Ravi B. Rapid tooling form accuracy estimation using region elimination adaptive search based sampling technique. Rapid Prototyping. 2007;13(1):182-190
7. Ollison T, Berisso K. Three-dimensional printing build variables that impact cylindricity. Journal of Industrial Technology. 2010;26(1):2-10
8. Lynn-Charney C, Rosen DW. Usage of accuracy models in stereolithography process planning. Rapid Prototyping Journal. 2000;6(2):77-86
9. Boschetto A, Bottini L. Design for manufacturing of surfaces to improve accuracy in fused deposition modeling. Robotics and Computer-Integrated Manufacturing. 2016;37:103-114
10. Béraud N, Vignat F, Villeneuve F, Dendievel R. New trajectories in electron beam melting manufacturing to reduce curling effect. Procedia CIRP. 2014;17:738-743
11. Zha W, Anand S. Geometric approaches to input file modification for part quality improvement in additive manufacturing. Journal of Manufacturing Processes. 2015;20:465-477

12. Pinto JM et al. Sensitivity analysis of geometric errors in additive manufacturing medical models. Medical Engineering and Physics. 2015;37:328-334

13. Szost BA et al. A comparative study of additive manufacturing techniques: Residual stress and microstructural analysis of CLAD and WAAM printed Ti–6Al–4V components. Materials & Design. 2015

14. Baufeld B, Brandl B, van der Biest O. Wire based additive layer manufacturing: Comparison of microstructure and mechanical properties of Ti–6Al–4V components fabricated by laser-beam deposition and shaped metal deposition. Journal of Materials Processing Technology. 2011;211:1146-1158

15. Shiva S et al. Investigations on the influence of composition in the development of Ni–Ti shape memory alloy using laser based additive manufacturing. Optics & Laser Technology. 2015;69:44-51

16. Cerniglia D, Scafidi M, Pantano A, Radlin J. Inspection of additive-manufactured layered components. Ultrasonics. 2015;62:292-298

17. Yang L, Harrison O, West H, Cormier D. Mechanical properties of 3D re-entrant honeycomb auxetic structures realized via additive manufacturing. International Journal of Solids and Structures. 2015;69–70:475-490

18. Yan C et al. Microstructure and mechanical properties of aluminium alloy cellular lattice structures manufactured by direct metal laser sintering. Materials Science and Engineering A. 2015;628:238-246

19. Karunakaran KP, Suryakumar S, Pushpa V, Akula S. Low cost integration of additive and subtractive processes for hybrid layered manufacturing. Robotics and Computer-Integrated Manufacturing. 2010;26:490-499

20. Sreenivasan R, Goel A, Bourrel DL. Sustainability issues in laser-based additive manufacturing. Physics Procedia. 2010;5:81-90

21. Wittbrodt BT et al. Life-cycle economic analysis of distributed manufacturing with open-source 3-D printers. Mechatronics. 2013;23:713-726

22. Achillas C et al. A methodological framework for the inclusion of modern additive manufacturing into the production portfolio of a focused factory. Journal of Manufacturing Systems. 2014

23. Mellor S, Hao L, Zhang D. Additive manufacturing: A framework for implementation. International Journal of Production Economics. 2014;149:194-201

24. Gao W et al. The status, challenges, and future of additive manufacturing in engineering. Computer-Aided Design. 2015

25. Conner BP et al. Making sense of 3-D printing: Creating a map of additive manufacturing products and services. Additive Manufacturing. 2014;1(4):64-76

26. Würtz G, Lasi H, Morar D. Additive manufacturing–enabling technology for lifecycle oriented value-increase or value-decrease. Procedia CIRP. 2015;33:394-399

27. Hedrick RW, Urbanic RJ, Burford CG. Development considerations for an additive manufacturing CAM system. IFAC-PapersOnLine. 2015;48(3):2327-2332

28. Weller C, Kleer R, Piller FT. Economic implications of 3D printing: Market structure models in light of additive manufacturing revisited. International Journal of Production Economics. 2015;164:43-56

29. Schröder M, Falk B, Schmitt R. Evaluation of cost structures of additive manufacturing processes using a new business model. Procedia CIRP. 2015;30:311-316

30. Baumers M, Dickens P, Tuck C, Hague R. The cost of additive manufacturing: Machine productivity, economies of scale and technology push. Technological Forecasting and Social Change. 2015

31. Zhang Y, Bernard A, Valenzuela JM, Karunakaran KP. Fast adaptive modeling method for build time estimation in additive manufacturing. CIRP Journal of Manufacturing Science and Technology. 2015;10:49-60

32. Huang R, Riddle M, Graziano D, Warren J, Das S, Nimbalkar S, Masanet E, et al. Energy and emissions saving potential of additive manufacturing: The case of lightweight aircraft components. Journal of Cleaner Production. 2016;135:1559-1570

33. Burkhart M, Aurich JC. Framework to predict the environmental impact of additive manufacturing in the life cycle of a commercial vehicle. Procedia CIRP. 2015;29:408-413

34. Newman ST, Zhu Z, Dhokia V, Shokrani A. Process planning for additive and subtractive manufacturing technologies. CIRP Annals - Manufacturing Technology. 2015;64:467-470

35. Fera M, Fruggiero F, Costabile G, Lambiase A, Pham DT. A new mixed production cost allocation model for additive manufacturing (MiProCAMAM). The International Journal of Advanced Manufacturing Technology. 2017;92(9–12):4275-4291

36. Costabile G, Fera M, Fruggiero F, Lambiase A, Pham D. Cost models of additive manufacturing: A literature review. International Journal of Industrial Engineering Computations. 2017;8(2):263-283

37. Fera M, Fruggiero F, Lambiase A, Macchiaroli R. State of the art of additive manufacturing: Review for tolerances, mechanical resistance and production costs. Cogent Engineering. 2016;3(1):1261503

38. Rickenbacher L, Spierings A, Wegener K. An integrated cost-model for selective laser melting (SLM). Rapid Prototyping Journal. 2013;19:208-214

39. Fera M, Macchiaroli R, Fruggiero F, Lambiase A. A new perspective for production process analysis using additive manufacturing—Complexity vs production volume. The International Journal of Advanced Manufacturing Technology. 2018;95(1–4):673-685

40. Fera M, Fruggiero F, Lambiase A, Macchiaroli R, Todisco V. A modified genetic algorithm for time and cost optimization of an additive manufacturing single-machine scheduling. International Journal of Industrial Engineering Computations, in press. 2018. DOI: 10.5267/j.ijiec.2018.1.001

41. Nearchou AC. Scheduling with controllable processing times and compression costs using population-based heuristics. International Journal of Production Research. 2010;48(23):7043-7062

CHAPTER-5

MODELING AND SIMULATION OF METAL AM

1. INTRODUCTION

1.1. Metal AM

Additive manufacturing (AM) has its roots in the development of stereolithography (SLA) in 1951 by Munz [1]. The development of SLA slowly escalated into a powder bed process and eventually into all of the various forms of AM that can be seen on the market today. In its infancy, the material and process controls only allowed AM to make prototypes and casting inserts. This was mainly due to the very low mechanical strength of the plastics being used and the process controls that produced very poor surface finishes [2]. This lacking drove researchers to find new and better materials and processes in order to use these methods for manufacturing as opposed to just prototyping. Over the last several decades, this has been accomplished and the American Society for Testing Materials (ASTM) has defined seven unique AM processes (binder jetting, directed energy deposition, material extrusion, material jetting, powder bed fusion, sheet lamination, and vat photopolymerization) [3]. The transition from rapid prototyping (RP) to AM resulted predominantly through the innovation of using metals in AM techniques. Of the seven standard AM process, three have been heavily investigated for their application to metal AM, powder bed fusion, directed energy deposition, and sheet lamination. In the process of sheet lamination, ultrasonic vibrations are used to fuse the sheets together. Due to its extreme dissimilarity to the other processes, namely the metal is never melted, it will be excluded from this body of work. The other processes, powder bed fusion and directed energy deposition, are so similar that many of the mathematical models used can be applied to both processes. To understand how these models are used it is best to understand the processes they simulate.

The first method to be developed for metal AM was powder bed fusion. In this process, shown in Figure 1a, the powder is delivered to the work area by a powder spreader of some type. The spreader is typically a roller, a pliable wiper, or a rigid wiper. This spreader will move the powder from the delivery system, either another bed or a hopper, to the work area. Once the powder is spread evenly across the surface of the work area, an energy source, typically a laser or electron beam, is used to selectively melt the powder into the desired solid shape. This process is repeated by lowering the work area and applying more powder with the spreader [4].

(a) Powder bed Fusion (b) Directed Energy Deposition (DED)

Figure 1. Schematics for most common metal AM processes.

The process of directed energy deposition, shown in Figure 1b, is radically different from the powder bed process. In this process an energy sources, again typically laser or electron beam, is used to melt a small portion of a substrate, sometimes called the build plate. Once a melt pool is generated, the material is fed into the melt pool. This material can take the form of either wire feedstock or powder, which is blown into the melt pool. The laser and feedstock are then traced along the surface of the substrate until the first layer is built. The machine is incremented in the z direction and the process is repeated. This is repeated until the entire part is built. As can be inferred, this process is not just limited to a 3 axis machine and more degrees of freedom can allow for more agility when building the part [5].

1.2. Modeling basics

In modeling a physical process, it is necessary to divide the space being modeled so that the equation can be computed. For AM processes, space representation has been divided into two main domain: Eulerian and Lagrangian. The Eulerian domain, which can also be called a field coordinate system, tracks specific locations in space to determine what is happening at a given location. This space, which can be seen in Figure 2a, is typically the space for finite element analysis (FEA). FEA can be used to simulate most physics processes. This process breaks the material down using a mesh structure and then the nodes are used to calculate the physics, which are being modeled. This mesh facilitates a quick and easy identification of neighbors using the connections of nodes found with the mesh. This technique is very beneficial when attempting to solve differential equations and has led to several solution methods being developed. However, since properties are typically saved as floating point values in the model, round off errors can lead to an arbitrary reduction or inflation of there properties.

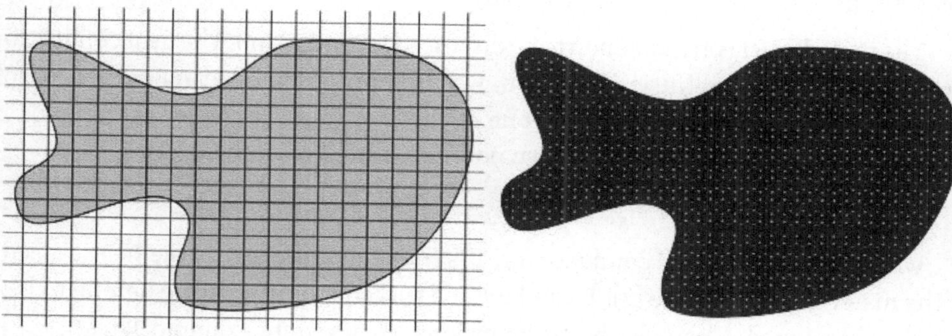

(a) Eulerian Domain (b) Lagrangian Domain

Figure 2. Depiction of domains used in metal AM modeling.

On the contrary, the Lagrangian domain, which can also be called a material coordinate system, tracks matter as it moves and reacts within the simulation space. In this domain, the material that is used will be broken up into discrete representative particles. Each of these particles is then tracked through space and time, calculating the various physical processes that apply to the particles. The Lagrangian method can be very beneficial for a couple of reasons. For starters, matter is never created nor destroyed by rounding of floating point numbers, due to the fact that the particle is given a fixed mass at the beginning of the simulation. In addition, the forces that the fluid is subject to are inherent within the simulation setup. Therefore, the pressure the fluid undergoes can be recorded without any additional calculations. And lastly, since particles are tracked there is no limit to where they are allowed to move. This can result in a much more accurate method of simulating fluid flow. This technique is not without its struggles. Since the material is divided into many little particles it can be very memory intensive and require substantial computing power. Additionally, this technique can make solving the differential equation that defines a system challenging due to the lack of inherent connectivity [6].

2. THERMAL MODELING

2.1. Modeling techniques

The heat transfer that occurs within a part can be broken down into three main types: conduction, convection, and radiation. Each of these can be calculated individually or the three can be combined to garner a full representation of the physical processes, which are taking place. For the sake of simplicity, each will be discussed on its own to highlight the necessary inputs and responses and then a complete picture will be drawn at the end.

The first physical phenomenon to be discussed is conduction. The heat conduction equation (or the heat diffusion equation) was first proposed by Jean Baptiste Joseph Fourier in 1807, resulting in it being one of the most mature modeling methods in AM. The full representation, in modern notation, can be seen in Eq. (1),

$$\nabla \cdot k\nabla T = c\frac{\delta T}{\delta t} \tag{1}$$

where k is the thermal conductivity, T is the temperature, c is the heat capacity of the material and t is time [7]. To understand this equation, it can be helpful to look at a simplified model, namely the 1-dimensional steady state example, Eq. (2),

$$q'' = -k\frac{dT}{dx} = k\frac{T_1 - T_2}{L} \tag{2}$$

where q" is the heat flux transferred within the part. From this representation of the equation, it can be seen that conduction is simply the material attempting to reach its lowest energy state, namely constant temperature. Though helpful for understanding, this approach applies to virtually no actual applications in AM modeling, for this, it is necessary to create a 3-dimensional transient model, shown in Eq. (3) [8].

$$\frac{\delta^2 T}{\delta x^2} + \frac{\delta^2 T}{\delta y^2} + \frac{\delta^2 T}{\delta z^2} = \frac{\rho c}{k}\frac{\delta T}{\delta t} \tag{3}$$

With the heat conduction within the part being handled by conduction, the other two phenomena only affect the surfaces of the material and are referred to as boundary conditions. The first, convection, is like conduction in the fact that utilizes contact to transfer heat from one object to another. Unlike conduction, which is between solid materials, convection is between a solid and a liquid, where air can be treated as a fluid. Newton, in 1701, saw that the relationship could be written as Eq. (4),

$$\frac{dT}{dt} \propto -(T - T_a) \tag{4}$$

where T is the temperature of the material, T a is the temperature of the fluid (*i.e.* ambient temperature), and t is time. The proportionally was later found for various situations and became known as the heat transfer coefficient (h) [9].

The last phenomena is a contact-less heat transfer through radiation. The law of radiation was first experimentally suggested by Josef Stefan in 1879 and later analytically shown by Ludwig Boltzmann in 1884, together these men derived the Stefan-Boltzmann Law, shown in Eq. (5),

$$q'' = \varepsilon\sigma T_s^4 \tag{5}$$

where q" is the heat flux due to radiation, ε is the emissivity of the surface, and σ is the Stefan-Boltzmann constant [10]. In order to implement these into a model, the process of conduction must be applied to the entirety of the material, whereas, the convection and radiation must only be applied to the skin as boundary conditions to ensure that the part cools properly based on the environment where it is located.

As can be seen from these equations, there are only five material properties, outline in Table 1, that are needed in order properly model the thermal history of an AM process. Most of these values must be found experimentally and therefore can be found in several locations including literature, material handbooks, and supplied by material vendors.

Table 1. Material properties required for thermal modeling.

Thermal conductivity (W / mK)
Specific heat (J / kgK)
Density (kg / m^3)
Convective coefficient (W / m^2K)
Emissivity

2.2. Validation techniques

Once the mathematical models have been developed it is necessary to show that they are a true representation of reality. For thermal modeling in AM, this has been done through two main methods, direct measurement of temperature with an instrument or by measuring a more easily attained physical characteristic and relating that to the temperature. As can be seen from a representative set of papers in Table 2, more effort has been placed on using an instrument to validate the model as opposed to measuring another characteristic. This is due to the fact that the physical characteristic comparison relies on the correctness of all models that are involved and does not have a one to one correlation between the material temperature and the model. Each of these techniques has a unique set of applications, strengths and concerns.

Table 2. Breakdown of validation techniques.

Instrument validated		Physical Char. validated	
IR/CCD camera	[11, 12, 13, 14]	Melt pool depth	[15, 16, 17]
Pyrometer	[18, 19]		
Thermal couple	[19, 20, 21]		

The use of a camera, be it IR or CCD, is a very appealing technique for many researchers for several reasons. The first key feature is that a camera is a non-contact measurement. This has several implications, namely, the application to all methods of metal AM developed to date and ability to be used in-situ. Since temperature is time and spatially dependent in nature, in order to accurately measure the temperature of a material both are critical, which in a camera is related to its frame rate and resolution. Researchers have reported frame rates of 800 frames/sec [11] and resolutions of 0.1 × 0.1 mm [12]. Though a powerful tool, cameras are not without their faults. One fault, related to the frame rate, is that a camera measures the average temperature of the skin during the time the shutter is open. This means that any extremely abrupt changes in the temperature can be smoothed by the camera. This problem only materializes when a pulsed laser is used, which has a different interaction with the material that includes an extremely fast and short spike in temperature [22]. The most critical consideration, however, is that cameras are highly sensitive to distance and angle to the substrate [23]. It is critical to ensure that both of these are controlled to ensure that an accurate measurement is made.

The next instrument used, a pyrometer, is less accurate than the camera but is simple which has allowed for it to be used as a feedback sensor where a mathematical model allows for the estimation of the expected sensor reading. A pyrometer is a non-contact spot instrument that is capable of measuring the average temperature of an area of the material. It measures the thermal radiation that is emitted from the area of the material where it is aimed. From Planck's radiation law and knowing the spectral distribution of the blackbody emissive power it is possible to derive Eq. (6) that is the effective temperature a pyrometer reads, assuming a Gaussian laser and a 1 mm radius for the pyrometer,

$$T_{eff} = \frac{hc}{\lambda\sigma \; ln\left(1 + n/\sum_{i=1}^{n}\left(e^{\frac{hc}{\lambda\sigma T_i}} - 1\right)^{-1}\right)} \tag{6}$$

where h is Planck's constant, c is the speed of light, λ is the wavelength of the emitted radiation, σ is Stefan-Boltzmann constant, n is the number of small sampling areas within the pyrometer viewing area, and T i is surface temperature within the small n areas [24]. This equation can then be included in a mathematical model to simulate the effect of a feedback system, which will give a much more accurate simulation.

The last instrument used for validation is a thermocouple, which is a contact spot measurement of the temperature. Thermocouples are limited to DED application because they must be fixed to the material. It also can not be used close to the melt pool or it will become detached and give inaccurate results. Lastly, it only is able to measure the average temperature of a specific location of the material, similar to the pyrometer. However, unlike the pyrometer that can be moved to follow the melt pool, thermocouples are fixed. This results in needing an array of thermocouples in order to garner a full understanding of the heat distribution within the material. For these reasons, thermocouples are typically used as a secondary validation.

The final validation technique is to use a more readily measured physical characteristic, namely melted track size, to show the correctness of the modeling approach. This method takes either a laser heating of the substrate or a single layer build and compares the results with the simulations findings. This is possible since the melting temperature of the material is known, therefore, the region of the simulation that reached a higher temperature can be flagged for analysis when the simulation is complete. The experiment is then sliced and the width and depth of the melted region are measured and then compared to the simulation. This method of validation is preferred by some due to the lack of needing any special equipment. However, the methods rely on the correctness of the fluid motion model, discussed in Section 3, and the thermal model previously discussed. This can lead to false rejection or validation of the model due to outside variables.

3. FLUID MOTION MODELING

Fluid modeling can be done in either the Eulerian domain or the Lagrangian domain. If the Eulerian domain is used the Volume of Fluid (VOF) technique is implemented. However, if a Lagrangian domain is used then either a Smoothed Particle Hydrodynamics (SPH) or Position Based Dynamics (PBD) can be implemented.

3.1. Volume of fluid

VOF is a method that is based on a Eulerian domain and tracks the amount of fluid that is within a specific region as a percentage of a fully dense region. In this method, the region being investigated is broken down into a mesh, like all Eulerian domain problems. Each mesh grid is then given an F value based on the amount of fluid that is within the cell. This F value denotes the percentage of the cell that is occupied by fluid; therefore, a value of 1 represents a cell that is full of fluid, a value of 0.5 represents a cell that is half full of fluid, and a value of 0 represents a cell that contains no fluid.

The definition of the F term allows for the fluid boundary to be readily determined by finding all cells with an F value between, but not equaling, 0 and 1. Calculating the gradient of F within the domain also allows for the normal of the fluid surface to be defined as the direction of most rapid change. It is possible to step the F value through time with Eq. (7).

$$\frac{\delta F}{\delta t}+u\frac{\delta F}{\delta x}+v\frac{\delta F}{\delta y}=0 \tag{7}$$

This is coupled with the Navier–Stokes equations, Eq. (8) and Eq. (9), and the continuity equation, Eq. (10), to find the motion that the fluid endures [25],

$$\frac{\delta(\rho u)}{\delta t}+\frac{\delta(\rho u^2)}{\delta x}+\frac{\delta(\rho uv)}{\delta y}=-\frac{\delta p}{\delta x}+g_x+\nu\left[\frac{\delta^2 u}{\delta x^2}+\frac{\delta^2 u}{\delta y^2}+\xi\left(\frac{1}{x}\frac{\delta u}{\delta x}-\frac{u}{x}\right)\right] \tag{8}$$

$$\frac{\delta(\rho v)}{\delta t}+\frac{\delta(\rho vu)}{\delta x}+\frac{\delta(\rho v^2)}{\delta y}=-\frac{\delta p}{\delta y}+g_y+\nu\left[\frac{\delta^2 v}{\delta x^2}+\frac{\delta^2 v}{\delta y^2}+\frac{\xi}{x}\frac{\delta v}{\delta x}\right] \tag{9}$$

$$\frac{1}{c^2}\frac{\delta p}{\delta t}+\frac{\delta u}{\delta x}+\frac{\delta v}{\delta y}+\frac{\xi u}{x}=0 \tag{10}$$

where u v is the velocity in the x y directions for Cartesian coordinate system or r z directions for cylindrical coordinate systems, ρ is the density, ξ is a coordinate specific variable 0 for Cartesian and 1 for cylindrical, g x g y is the body acceleration, ν is the kinematic viscosity, p is the pressure, t is time, and c is the adiabatic speed of sound in the fluid.

In order to step through time, the following steps must be followed:

- ☆ Compute the first guess of the new velocities with explicit approximations of Eq. (8) and Eq. (9) using previous time step values as the boundary conditions.
- ☆ Adjust the pressure in each cell to satisfy Eq. (10) and change velocities of adjacent cells accordingly. Iterate until pressures and velocities are found such that all equations are balanced.
- ☆ Update F with Eq. (7) from the solved values

In order to model with the VOF techniques, the material properties outlined in Table 3 need to be known. Similar to heat transfer material properties, these properties must be found experimentally and can be found throughout literature and material properties handbooks.

Table 3. Material properties required for VOF fluid modeling.

Density (kg / m^3)
Kinematic viscosity ($P_{a.s}$)

Though a mature and robust method, one of the main disadvantages of this method is its reliance on a Eulerian domain. This results in the fluid mass in a cell containing a free surface being represented by a floating point number. In a computer system, the precision is limited, which will lead to round off errors. In a technique as VOF, this can lead to loss or creation of mass based on the rounding of these numbers. This, in turn, breaks the law of conversation of mass. Another problem that arises from the use of VOF techniques is that the fluid is only allowed to move within predetermined space becuase the technique relies on the eularian domain. Even if some of the fluid physically would move outside this domain, this model will not allow this resulting in the potential for inaccurate results.

3.2. Smoothed particle hydrodynamics

The SPH method utilizes a Lagrangian domain and approximates the fluid as a set of particles that are tracked through space. This method is based on the fact that a function f x can be written with an integral representation, Eq. (11),

$$f(x) = \int_{\Omega} f(x')\delta(x - x')dx' \tag{11}$$

where Ω is the domain and δ is a Dirac function. For the SPH technique, it is then necessary to convert from a continuous function to a discrete function because the fluid is represented by a set of particles and not a continuous fluid. Additionally, the Dirac function can be replaced with a weighting function to result in Eq. (12),

$$f(x) = \sum_{b=1}^{N} f(x_b)W(x - x_b, h)\Delta V_b \tag{12}$$

where b are the neighboring particles, W is the weighting function, and h is the radius of influence of the weighting kernel.

$$W(r,h) = \alpha_d e^{-(r/h)^2} \tag{13}$$

$$W(r,h) = \alpha_d \begin{cases} \left(1 + 3\frac{r}{h}\right)\left(1 - \frac{r}{h}\right)^2 & \frac{r}{h} \leq 1 \\ 0 & \frac{r}{h} > 1 \end{cases} \tag{14}$$

$$W(r,h) = \alpha_d \begin{cases} \frac{3}{2} - \left(\frac{r}{h}\right)^2 + \frac{1}{2}\left(\frac{r}{h}\right)^3 & 0 \leq \frac{r}{h} < 1 \\ \frac{1}{6}\left(1 - \frac{r}{h}\right)^3 & 1 \leq \frac{r}{h} < 2 \\ 0 & \frac{r}{h} \geq 2 \end{cases} \tag{15}$$

The weighting function acts as a smoothing function that gives a weight to the influence that the particles have on each other. This weighting function results in particles that are closer together having more influence than those that are farther away. The various weighting kernels have been proposed in literature including Gaussian Kernel Eq. (13), Bell shaped kernel Eq. (14), or Cubic spline kernel Eq. (15), where r is the distance between the particles begin investigated and α d is a normalizing scalar.

The conservation of mass, Eq. (16), within this system is the first equation that needs to be written in this approximates form, Eq. (17)

$$\frac{D\rho}{Dt} = u\nabla\rho - \nabla(\rho u) \tag{16}$$

$$\frac{D\rho_a}{Dt} = \sum_{b=1}^{N} m_b u_{ab} \nabla_a W_{ab} \tag{17}$$

Next, conservation of momentum, Eq. (18), must be written in it's approximate form, Eq. (19).

$$\frac{Du}{Dt} = -\frac{1}{\rho}\nabla p \tag{18}$$

$$\frac{Du_a}{Dt} = -\sum_{b=1}^{N} m_b \left(\frac{p_b}{\rho_b^2} + \frac{p_a}{\rho_a^2} + \Pi_{ab}\right)\nabla_a W_{ab} + F \tag{19}$$

where the artificial viscosity Π ab is defined by Eq. (20),

$$\Pi_{ab} = \begin{cases} \dfrac{-\alpha c_{ab}\mu_{ab} + \beta\mu_{ab}^2}{\rho_{ab}} & u_{ab}(x_a - x_b) < 0 \\ 0 & u_{ab}(x_a - x_b) \geq 0 \end{cases} \tag{20}$$

where c ab is the average speed of sound of the particles, ρ ab is the average density of the particles, and μ is defined in Eq. (21).

$$\mu_{ab} = \frac{h u_{ab}(x_a - x_b)}{(x_a - x_b)^2 + (0.01h^2)^2} \tag{21}$$

The volume forces, F , is any force that is applied to the entire fluid, such as gravity. One final equation that is needed is the equation of state, which links the pressure and density, for an ideal gas it can be defined as Eq. (22) or for a quasi-incompressible fluid it can be defined as Eq. (23),

$$p = \frac{RT}{M}\left(\frac{\rho}{\rho_0} - 1\right) \tag{22}$$

$$p = c_0^2\rho_0\left(\frac{\rho}{\rho_0} - 1\right) \tag{23}$$

where R is the ideal gas constant, T is the temperature, M molar mass, c 0 is the speed of sound in the material [26].

In order to step through time, the following steps must be followed for each particle:

- ☆ Find all neighbors within smoothing kernel length
- ☆ Solve Momentum Equation by solving for Artificial Viscosity (Eq. (20)) and Equation of State (Eq. (22) or Eq. (23)) and applying Volume Forces
- ☆ Solve Continuity Equation (Eq. (17))
- ☆ Update ρ, *x*, *u*

Even though it has some very appealing properties, SPH is not without its faults. The main fault applying to AM modeling is that the SPH technique is very sensitive to density fluctuations resulting from a lack of neighbors. In AM simulation, this problem can arise when a simulation is beginning to develop a melt pool, when a melt pool is in the end stages of solidification, or if the energy source is moving so fast that only a small amount of material is melted. This fluctuation in density can be combated with extremely small time steps, however, this makes the computation time extremely long [27].

3.3. Position based dynamics

PBD is also a Lagrangian particle based method, similar to SPH. This method approximates the domain with particles that are subject to the standard Newtonian forces. These forces are applied to all of the particles and their position in space is tracked to determine the motion of the fluids.

In order to solve for the position of the particle sets, several functions must be applied to the particles. The first function that needs to be defined is the scheme for updating the velocities of the set of particles. This is done by adjusting the velocities based on any external forces, Eq. (24),

$$v_i^* = v_i + \Delta t f_{ext} \tag{24}$$

where v i ∗ is the new velocity of the particle, v i is the current velocity of the particle, Δ t is the time step, and f ext is the sum of the external forces that are applied to the particle. From this, a guess is made as to the new positions of the particles, Eq. (25).

$$x_i^* = x_i + \Delta t v_i \tag{25}$$

Next, it is necessary to define the equations that enforce incompressibility of the fluid. The first equation is a function of the position (p 1 ... p n) of all of the neighbors, Eq. (26),

$$C_i(p_1 \dots p_n) = \frac{\rho_i}{\rho_0} - 1 \tag{26}$$

where ρ 0 is the rest material density and ρ i is the standard SPH density Eq. (19). Since the density of the fluid is assumed to be constant, it is possible to find the pressure change each particle experiences, Eq. (27),

$$\Delta p \approx \nabla C(p)\lambda \tag{27}$$

where λ is a scalar. From this, it is possible to solve for λ , Eq. (28)

$$\lambda_i = -\frac{C_i(p_1 \dots P_n)}{\sum_k \left|\nabla_{p_k} C_i\right|^2} \tag{28}$$

Once λ is known, it is then possible to determine the change in pressure that each particle experiences with Eq. (29)

$$\Delta p_i = \frac{1}{\rho_0} \sum_j (\lambda_i + \lambda_j) \nabla W\left(p_i - p_j, h\right) \tag{29}$$

Once the pressure change is known, it is then possible to generate the actual positions (x i ∗) of the particles with Eq. (30).

$$x_i^* = x_i + \Delta p_i \tag{30}$$

And lastly, with the actual motion of the particles known, it is possible to determine the final velocity of the particles from Eq. (31).

$$v_i^* = \frac{1}{\Delta t}(x_i^* - x_i) \tag{31}$$

To step through time each of these steps is to be applied to each particle in the simulation before moving on to the next step [27]:

- ☆ Apply external forces (Eq. (24)) and predict the position (Eq. (25))
- ☆ Find the nearest neighbors
- ☆ Calculate the λ i (Eq. (28))
- ☆ Calculate change in pressure (Eq. (29)) and collision detection
- ☆ Update positions (Eq. (30)) and velocities (Eq. (31))

Like SPH, this method is appealing due to its inherent conservation of mass and Lagrangian representation does not limit the flow to a specific area. However, this technique relies on arbitrary values to generate an accurate appearing simulation. For this reason, tuning is needed to develop a set of inputs, as opposed to running an experiment to gather the necessary material properties.

4. STRESS MODELING

Once the evolution of heat through the process has been found, it is then possible to take this data and estimate the stresses and deformations that are the result of the cyclic process.

4.1. Modeling techniques

The total stress that accumulates during an AM process can be written as the sum of the individual contributors of the strain, Eq. (32),

$$\varepsilon_{ij} = \varepsilon_{ij}^{E} + \varepsilon_{ij}^{P} + \varepsilon_{ij}^{T} + \varepsilon_{ij}^{\Delta V} + \varepsilon_{ij}^{TrP} \tag{32}$$

where ε ij is the total strain, ε ij E is elastic strain (Eq. (33)), ε ij P is the plastic strain (Eq. (34)), ε ij T is thermal strain (Eq. (35)), ε ij Δ V is strain from volumetric dilation strain (Eq. (36)), and ε ij TrP is the strain from phase transformation (Eq. (37)).

$$\varepsilon_{ij}^{E} = \frac{1+\nu}{E}\sigma_{ij} - \frac{\nu}{E}\sigma_{kk}\delta_{ij} \tag{33}$$

$$\varepsilon_{ij}^{P} = \widehat{G}\left(\frac{\delta F}{\delta \sigma_{kl}} + \frac{\delta F}{\delta T}T + \sum \frac{\delta F}{\delta X_i}X_i\right)\frac{\delta F}{\delta \sigma_{ij}} \tag{34}$$

$$\varepsilon_{ij}^{T} = \alpha(T - T_0)\delta_{ij} \tag{35}$$

$$\varepsilon_{ij}^{\Delta V} = \frac{\Delta V}{3V}\Delta X \tag{36}$$

$$\varepsilon_{ij}^{TrP} = \frac{5S_{ij}\Delta V}{4YV}(2 - 2X_n - \Delta X)\Delta X \tag{37}$$

where E is Young's modulus, ν is Poisson's ratio, α is the thermal expansion coefficient, Y is the yield stress of the weaker phase, F is the yield function (Eq. (38)) and G is the hardening function (Eq. (39)) [28].

$$F = F(\sigma_{ij}, \varepsilon^{P}, \kappa, T, X_i) \tag{38}$$

$$\frac{1}{\widehat{G}} = -\left(\frac{\delta F}{\delta \sigma_{mn}^{P}} + \frac{\delta F}{\delta \kappa}\sigma_{mn}\right)\frac{\delta F}{\delta \sigma_{mn}} \tag{39}$$

In order to solve for the stress that is accompanied by the AM process, each of the individual components of strain need to be solved for and added together. This can be a tedious and computationally expensive process, based on the parameters in Table 4.

Table 4. Material properties required for stress modeling.

Young's modulus
Poisson's ratio
Thermal expansion coefficient
Yield stress of the weaker phase

To reduce the simulation time, the inherent strain approach has been developed. In this method the strains are lumped together, as seen in Eq. (40), to simplify the calculations.

$$\varepsilon^* = \varepsilon_{ij} - \varepsilon_{ij}^E = \varepsilon_{ij}^P + \varepsilon_{ij}^T + \varepsilon_{ij}^{\Delta V} + \varepsilon_{ij}^{TrP} \tag{40}$$

This technique will run a small scale simulation to determine the value of this inherent strain. These values are then superimposed as needed when the full-scale simulation is being modeled. This results in a simulation that is orders of magnitude faster than the previously described technique [29].

4.2. Validation techniques

Once the models have been developed it is necessary to validate the technique. There are several methods that have been used, which are reported in Table 5.

Table 5. Frequency of stress analysis techniques.

Presence of cracks	[13, 30]
Final part displacement	[31, 32, 21]
Digital image correlation (DIC)	[33, 34]
Neutron diffraction	[35, 20]
X-ray diffraction	[36, 37]

The simplest of the methods that have been used is to observe the locations of any cracks that have developed during the build. The experimental crack locations are then compared to the simulation to determine if a high stress is predicted. This method is appealing due to the lack of specialized equipment required. However, the lack of detail results in this method only being able to be used as a method of qualitative validation and not a precise comparison.

A more precise comparison is needed in order to quantitatively compare the simulation and the experiment, and this has led to the following techniques. The first technique is to measure the distortion that occurs within the part during the build process. This has been done with either a laser displacement sensor or a 3-D scanner. If a laser displacement sensor is to be used then the substrate must be setup as a cantilever, as in Figure 3, which will allow one end to move as the material is heated and cools. The fluctuations that are observed in the end of the substrate are then compared to the simulation to determine the accuracy of the simulation. Another method of measuring the distortion of the part is to scan the part with a 3-D scanner after the part has been removed from the fixturing and allowed to flex. Each of these methods has its advantages and problems. The first is the accuracy of the validation technique. For a higher accuracy, the laser displacement sensor should be used, and it has been reported to have an accuracy of ± 1 μm [21], whereas 3D scanners have reported to have an accuracy of ± 500 μm [32]. Coupled with the accuracy is the area that is measured to determine the distortion. The laser displacement sensor only measures a single spot and tracks that through space and time, whereas the 3D scanner is capable of capturing the distortion that occurred in a larger section of the build. An additional advantage that the 3-D scanner has over the laser displacement sensor is the effect of misalignment of the initial deposit is less when compared to the laser displacement sensor. However, the 3D scanner cannot be used in-situ and the laser displacement sensor can. These methods allow for a quantitative analysis of the distortion which can be used to determine the stress within the part.

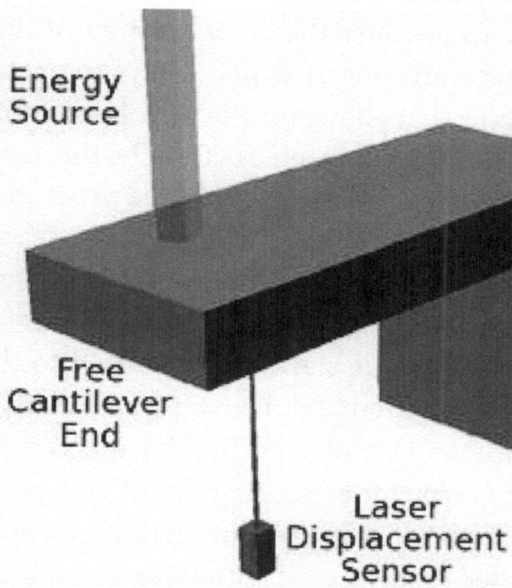

Figure 3. Experimental setup using laser displacement sensor to measure distortion.

Another method of tracking the part's distortion is to use digital image correlation (DIC). This method uses a camera to observe the part and track any distortion by comparing frames. Points are selected on the material and tracked through space as shown in Figure 4.

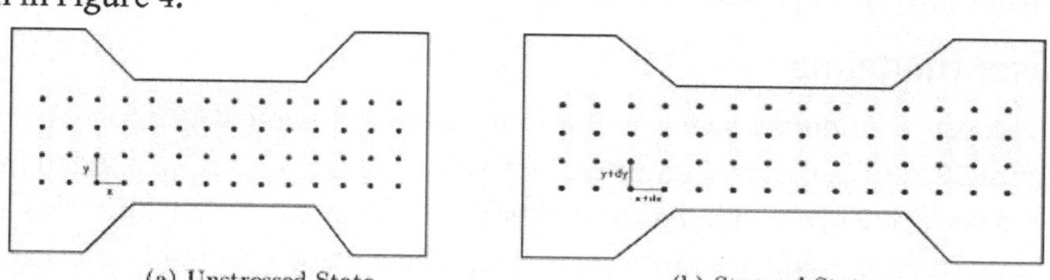

Figure 4. Schematic of mini-tensile specimen with tracking points for DIC.

This method will inherently return the distortion that a part undergoes and it then it is possible to precisely determine any surface level stresses. Measuring the stresses is possible by starting with a stressed specimen, for example, one that has undergone the deposition process. The part is then selectively stress relieved to determine the relaxation that the material undergoes. The relaxation that the material experiences will be directly linked to the stress [33]. This method is highly appealing because the motion of the material is measured in pixels of the camera. Because of this, the resolution of this method is limited by only the camera, which can be increased by using a microscope in conjunction with a camera. The major drawback of this technique is that it is a destructive method, resulting in it not being an acceptable method for some applications.

If the exact strain, the displacement of the atoms from their rest positions, is needed then it is necessary to measure the actual motion of the atoms. This is done using Bragg's Law and the scattering of either X-Rays or neutrons. The initial material is placed in an apparatus that allows for the initial diffraction pattern to be generated. This is the initial location of the atoms before the thermal processes have been applied to the material. The material is then subjected to the thermal processes. The material is then placed back into the apparatus and the diffraction pattern is again measured. This new pattern is the final locations of the atoms. From these two sets, it is possible to find how far the atoms moved from their base locations, this value is the strain. This strain value can then be related to the simulation to determine the accuracy of the simulation [38]. Though an accurate method, the use diffraction patterns are not without its faults. The first is the access to researchers, it is necessary to have a dedicated setup for measuring the diffraction, which can make it difficult for some to gain access. Additionally, this method can only be used at the end of a build and cannot be used to measure in-situ strain.

5. MISCELLANEOUS MODELS

In order to completely model the AM process it can be necessary to include an energy source, laser or e-beam, and if powders are used it can be necessary to track their position within a powder bed or blown powder.

5.1. Laser modeling

The most common laser distributions used in metal AM process are top hat, Figure 5b, and Gaussian, Figure 5a. Just as the material must be divided into the domain, a laser must be divided into smaller pieces for proper modeling. This is done by dividing the projected area into segments, which can be thought of as rays from the laser. The flux that is generated within each of these divisions of the laser is found by taking the integral of the functions over one division. This was done for the Gaussian laser distribution where Figure 5a was generated by integrating Eq. (41),

$$\phi(x,y) = \phi_0\sqrt{1 - \frac{x^2}{r_0^2} - \frac{y^2}{r_0^2}} \tag{41}$$

where φ_{xy} is the intensity as a specific $x\,y$ location, φ_0 is the initial laser power, and r_0 is the laser radius [19]. The top hat beam, Figure 5b, is much easier to model. In this model, the laser power is simply divided by the number of divisions that can be placed within the laser boundaries. The heat flux that is found is then applied to a specific location of the domain based on the current laser location. These intensities are then multiplied by α, the material absorptivity, to determine the actual amount of energy that is absorbed by the material. This process can be applied to any distribution that can be imagined.

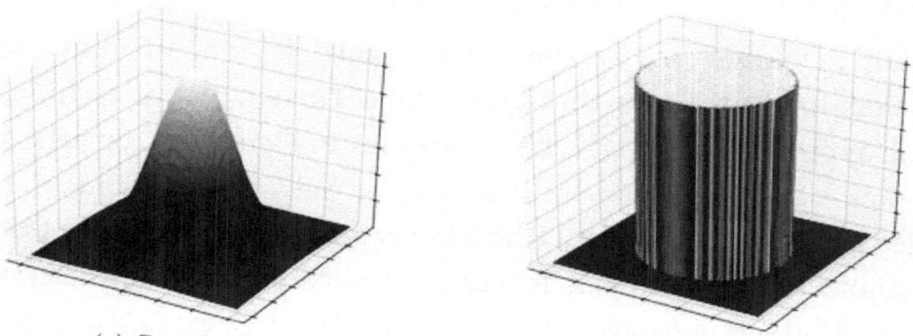

(a) Gaussian Laser (b) Top Hat Laser

Figure 5. Example of heat flux from example distributions.

There are two methods of applying the laser. The first is to define the surface and apply the flux directly to this region of the material and not attempting to determine if there is anything blocking the laser projection. The other, and more realistic method, is to perform ray tracing and apply the heat flux to the first location that is struck by the ray. Both method work but ray tracing is preferred for the sake of reality but can be costly to compute.

5.2. E-beam modeling

The other choice of a heat source for metal AM is an electron beam. Typically an electron beam is modeled as a Gaussian heat source, just as the previously mentioned laser. However, an electron beam will also penetrate the surface heating the material for a given depth. The intensity in the z-direction can be expressed as Eq. (42).

$$\phi_z = \frac{1}{0.75}\left(-2.25\left(\frac{z}{S}\right)^2 + 1.5\frac{z}{S} + 0.75\right) \tag{42}$$

The xy and z intensities can then be combined to determine the heat flux, as shown in Eq. (43),

$$\dot{Q}(x,y,z) = -\eta_b\eta_e P_B\frac{\phi_{xy}\phi_z}{S} = -\eta_b\eta_e P_B\frac{4\ln(0.1)}{\pi d_b^2 S}e^{\frac{4\ln(0.1)(x^2+y^2)}{d_b^2}}\left(-3\left(\frac{z}{S}\right)^2 - 2\frac{z}{S} + 1\right) \tag{43}$$

where η_b is the beam control efficiency, η_e is the energy conservation at the part surface, P_B is the beam power, S is the absolute penetration depth, and d_b beam diameter [39].

5.3. Powder bed and blown powder models

The last element of the metal AM process that is typically necessary is the addition of material. In the wire feed DED process this is done by modeling the wire as a solid material, just as the substrate, and treating it in a similar fashion. When using either a powder bed or blown powder, this is not possible. Due to the stochastic nature of the powders, it can be challenging to model their behavior. When modeling the powder bed setup, there are two prevailing methods that have been used: discrete element method (DEM) and geometric methods.

The DEM technique tracks the powder particles on an individual basis to determine their final position in the build volume. This simulation technique typically begins by dropping particles (the blue particles in Figure 6a), or sets of particles, from a random x and y position but a designated height in the domain. From there, the particles position and velocity (the red vectors in Figure 6a) are tracked as the particle is subjected to the major forces of gravity, contact, and friction. In some cases, smaller interaction forces, such as Van der Waal forces [40] or JKR interaction forces [41], are added to increase the accuracy of the simulation. The particles are allowed settle to their resting point (the green particles in Figure 6a) and more particles are added as needed to the simulation. This technique is very appealing to the simulation powder spreading in powder bed AM due to the ease of adding a powder spreader to the simulation without much additional effort. This results in the ability to simulation the entire process from the layering of the powder to laser interaction to melting and solidification [42].

The geometric method is not as realistic for powder bed process but is computationally only a small fraction of the DEM technique. In this technique, the area to be filled is analyzed without regard to how the powder would flow. The first geometric method is referred to as the compression algorithm. In this technique, the particles are randomly spaced within the domain. The particles are then moved in the direction of compression, typically the direction of gravity, by the shortest distance that results in a collision of particles. This compressing is repeated until the potential energy is below a user-defined tolerance. The particles are then "shaken", or moved laterally, and the process is repeated. The build volume is then refilled with more particles and the process is repeated until the volume is full [43]. Another geometric method, displayed in Figure 6b, focuses on a tetrahedral mesh. In this method, the build volume is meshed and particles are placed on the nodes and edges based on a set of rules. This allows for an efficient filling of the space that has a packing density approximately matching reality [44].

Each of these methods can be used to simulate the powder bed process, however, the selection can be based on a couple of main factors. If speed is required then a geometric approach should be used. This approach takes on the order of seconds for a full simulation whereas the DEM approach will take hours to days to simulation an identical setup. If physical accuracy is needed then it necessary to use the DEM approach. This is because it tracks the powder particles through time and space. They are subjected to the forces of nature that result in a realistic simulation, whereas the geometric approach is simply a packing problem where the particles are placed where they can fit. This can result in packing densities that are not representatives of natural occurrences.

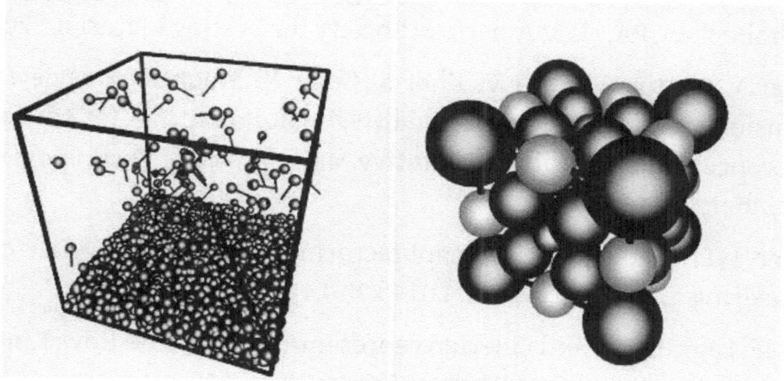

Figure 6. Powder bed modeling techniques.

Modeling of the blown powder DED has only been done with the DEM approach previously discussed. To apply the DEM to blown powder, the nozzle must be modeled as a boundary condition and the particles should be fed through the feed system and tracked to determine when and where they strike the melt pool.

6. CONCLUSION

In order to mathematically model the AM process, it is necessary to couple several distinct mathematical models. The necessary models are thermal, fluids modeling and energy input modeling. Other models that can be included, based on the user's desires, are stress, microstructure, surface finish, and more attributes.

Some examples where this have been done are:

- ☆ Fan and Liou [45] model heat transfer and fluid flow dynamics (VOF) for a blown powder DED AM method with a laser for Ti-64
- ☆ Kumar et al. [46] model heat transfer and fluid flow dynamics (SPH) for a wire feed DED AM method with a laser for Ti-64
- ☆ Lee and Zhang [47] model powder bed generation (DEM), heat transfer, fluid flow (VOF) and microstructure for powder bed AM with Nickel-based superalloys

REFERENCES

1. Munz OJ. Photo-Glyph Recording; 1951
2. Bourell DLD, Beaman JJ, Leu MC, Rosen DWA. Brief history of additive manufacturing and the 2009 roadmap for additive manufacturing: Looking back and looking ahead. In: Workshop on Rapid Technologies. 2009. pp. 5-11. Available from: http://iweb.tntech.edu/rrpl/rapidtech2009/bourell.pdf
3. ASTM. Standard Guidelines for Design for Additive Manufacturing. West Conshohocken, PA, USA: American Society for Testing Materials; 2017. p. 52910
4. Bhavar V, Kattire P, Patil V, Khot S, Gujar K, Singh R. A review on powder bed fusion technology of metal additive manufacturing. In: 4th International Conference and Exhibition on Additive Manufacturing Technologies-AM-2014, September; 2014
5. Frazier WE. Metal additive manufacturing: A review. Journal of Materials Engineering and Performance. 2014;23(6):1917-1928
6. Price JF. Lagrangian and Eulerian representations of fluid flow: Kinematics and the equations of motion. MIT OpenCourseWare; 2006
7. Narasimhan TN. Fourier's heat conduction equation: History, influence, and connections. Reviews of Geophysics. 1999;31(1):151-172
8. Han JC. Analytical Heat Transfer. Boca Raton, FL: CRC Press; 2016
9. Besson U. The history of the cooling law: When the search for simplicity can be an obstacle. Science & Education. 2010 Nov;21(8):1085-1110. DOI: 10.1007/s11191-010-9324-1
10. Crepeau J. A Brief History of the T 4 Radiation Law; 2009. pp. 59-65
11. Hu D, Kovacevic R. Sensing, modeling and control for laser-based additive manufacturing. International Journal of Machine Tools and Manufacture. 2003;43:51-60
12. Kolossov S, Boillat E, Glardon R, Fischer P, Locher M. 3D FE simulation for temperature evolution in the selective laser sintering process. International Journal of Machine Tools and Manufacture. 2004;44:117-123
13. Zhu G, Zhang A, Li D, Tang Y, Tong Z, Lu Q. Numerical simulation of thermal behavior during laser direct metal deposition. The International Journal of Advanced Manufacturing Technology. 2011;55(9–12):945-954. Available from: http://link.springer.com/10.1007/s00170-010-3142-0

14. Zeng K, Pal D, Stucker BE. A Review of thermal analysis methods in laser sintering and selective laser melting. In: Proceedings of the Solid Freeform Fabrication Symposium;2012. pp. 796–814. Available from: http://utwired.engr.utexas.edu/lff/symposium/proceedingsArchive/pubs/Manuscripts/2012/2012-60-Zeng.pdf

15. Contuzzi N, Campanelli SL, Ludovico AD. 3D finite element analysis in the selective laser melting process. International Journal of Simulation Modelling. 2011;10(3):113-121

16. Khairallah SA, Anderson A. Mesoscopic simulation model of selective laser melting of stainless steel powder. Journal of Materials Processing Technology. 2014;214(11)

17. Tang Q, Pang S, Chen B, Suo H, Zhou J. A three dimensional transient model for heat transfer and fluid flow of weld pool during electron beam freeform fabrication of Ti-6-Al-4-V alloy. International Journal of Heat and Mass Transfer. 2014;78:203-215. DOI: 10.1016/j.ijheatmasstransfer.2014.06.048

18. Dai K, Li X, Shaw LL. Comparisons between thermal modeling and experiments: Effects of substrate preheating. Rapid Prototyping Journal. 2004;10(1):24-34. Available from: http://www.emeraldinsight.com/doi/10.1108/13552540410512507

19. Peyre P, Aubry P, Fabbro R, Neveu R, Longuet A. Analytical and numerical modelling of the direct metal deposition laser process. Journal of Physics D: Applied Physics. 2008;41(2):025403. Available from: http://stacks.iop.org/0022-3727/41/i=2/a=025403?key=crossref.1de7d31af0ef9a95b456f9aa3d8caf90

20. Ding J, Colegrove P, Mehnen J, Ganguly S, Almeida PMS, Wang F, et al. Thermo-mechanical analysis of wire and arc additive layer manufacturing process on large multi-layer parts. Computational Materials Science. 2011;50(12):3315-3322. DOI: 10.1016/j.commatsci.2011.06.023

21. Heigel JC, Michaleris P, Reutzel EW. Thermo-mechanical model development and validation of directed energy deposition additive manufacturing of Ti-6Al-4V. Additive Manufacturing. 2015;5:9-19. DOI: 10.1016/j.addma.2014.10.003

22. Fischer P, Locher M, Romano V, Weber HP, Kolossov S, Glardon R. Temperature measurements during selective laser sintering of titanium powder. International Journal of Machine Tools and Manufacture. 2004;44(12–13):1293-1296

23. Wegner A, Witt G. Process monitoring in laser sintering using thermal imaging. In: Proceedings of the Solid Freeform Fabrication Symposium;

2011. pp. 405-414. Available from: http://www.scopus.com/inward/record.url?eid=2-s2.0-84898417627

24. Zhang Y. Thermal modeling of advanced manufacturing technologies: Grinding, laser drilling, and solid freeform fabrication [Ph.D. thesis]. Ann Arbor, MI: University of Connecticut; 1998. Available from: http://opencommons.uconn.edu/dissertations/AAI9918103

25. Hirt CW, Nichols BD. Volume of fluid (VOF) method for the dynamics of free boundaries. Journal of Computational Physics. 1981;39:201-225

26. Goffin L. Development of a didactic SPH model [Masters thesis]; 2013

27. Macklin M, Müller M. Position based fluids. ACM Transactions on Graphics (TOG). 2013;32(4):104

28. Ghosh S, Choi J. Modeling and experimental verification of transient/residual stresses and microstructure formation in multi-layer laser aided DMD process. Journal of Heat Transfer. 2006;128(7):662. Available from: http://heattransfer.asmedigitalcollection.asme.org/article.aspx?articleid=1448394

29. Murakawa H, Deng D, Ma N, Wang J. Applications of inherent strain and interface element to simulation of welding deformation in thin plate structures. Computational Materials Science. 2012;51(1):43-52. DOI: 10.1016/j.commatsci.2011.06.040

30. Gusarov AV, Pavlov M, Smurov I. Residual stresses at laser surface remelting and additive manufacturing. Physics Procedia. 2011;12:248-254

31. Liu H, Sparks TE, Liou FW. Numerical analysis of thermal stress and deformation in multi layer laser metal deposition process. In: Proceedings of the Solid Freeform Fabrication Symposium; 2013. p. 577-591

32. Denlinger ER, Irwin J, Michaleris P. Thermomechanical modeling of additive manufacturing large parts. Journal of Manufacturing Science and Engineering. 2014;136(6):061007. Available from: http://manufacturingscience.asmedigitalcollection.asme.org/article.aspx?doi=10.1115/1.4028669

33. Wu AS, Brown DW, Kumar M, Gallegos GF, King WE. An experimental investigation into additive manufacturing-induced residual stresses in 316L stainless steel. Metallurgical and Materials Transactions A: Physical Metallurgy and Materials Science. 2014;45(13):6260-6270

34. King W, Anderson AT, Ferencz RM, Hodge NE, Kamath C, Khairallah SA. Overview of modelling and simulation of metal powder bed fusion process at Lawrence Livermore National Laboratory. Materials Science and Technology. 2015;31(8):957-968. Available from: http://www.tandfonline.com/doi/full/10.1179/1743284714Y.0000000728

35. Ding J, Colegrove P, Mehnen J, Williams S, Wang F, Almeida PS. A computationally efficient finite element model of wire and arc additive manufacture. International Journal of Advanced Manufacturing Technology. 2014;70(1–4):227-236

36. Yadroitsava I, Yadroitsev I. Residual stress in metal specimens produced by direct metal laser sintering. In: Proceedings of the Solid Freeform Fabrication Symposium; 2013. pp. 1689–1699

37. Shah K, Haq I, Shah SA, Khan F, Khan MT, Khan S. Experimental study of direct laser deposition of Ti-6Al-4V and inconel 718 by using pulsed parameters. The Scientific World Journal. 2014;2014:6

38. Fitzpatrick ME, Fry AT, Holdway P, Kandil FA, Shackleton J, Suominen L. Determination of residual stresses by X-ray diffraction–Issue 2. Measurement Good Practice Guide. 2005;52:74

39. Zäh MF, Lutzmann S. Modelling and simulation of electron beam melting. Production Engineering. 2010;4(1):15-23

40. Cheng YF, Guo SJ, Lai HY. Dynamic simulation of random packing of spherical particles. Powder Technology. 2000;107:123-130

41. Johnson KL, Kendall K, Roberts AD. Surface Energy and the Contact of Elastic Solids. Vol. 324; 1971. pp. 310-313

42. Yang RY, Zou RP, Yu AB. Computer simulation of the packing. Physical Review E. 2000;62(3):3900-3908

43. Han K, Feng YT, Owen DR. Sphere packing with a geometric based compression algorithm. Powder Technology. 2005;155:33-41

44. Jerier JF, Richefeu V, Imbault D, Donze F. Packing spherical discrete elements for large scale simulations. Computer Methods in Applied Mechanics and Engineering. 2010;199:1668-1676

45. Fan Z, Liou F. Numerical modeling of the additive manufacturing (AM) processes of titanium alloy. In: Titanium Alloys–Towards Achieving Enhanced Properties for Diversified Applications. Croatia, Rijeka: InTech; 2012. pp. 3-29. Available from: http://www.intechopen.com/books/titanium-alloys-towards-achieving-enhanced-properties-for-diversified-applications/numerical-modeling-of-the-additive-manufacturing-am-processes-of-titanium-alloys

46. Kumar KS, Sparks TE, Liou F. Parameter determination and experimental validation of a wire feed additive manufacturing model. In: Solid Freeform Fabrication Symposium. Vol. 1. 2015. pp. 1129-1153

47. Lee YS, Zhang W. Modeling of heat transfer, fluid flow and solidification microstructure of nickel-base superalloy fabricated by laser powder bed fusion. Additive Manufacturing. 2016;12:178-188. DOI: 10.1016/j.addma.2016.05.003

Part-2

3D PRINTING AND MEDICINE

CHAPTER-6

PATIENT-SPECIFIC 3D PRINTED MODELS FOR EDUCATION, RESEARCH AND SURGICAL SIMULATION

1. INTRODUCTION

Within recent years there has been exponential growth in healthcare related 3D printing research (as shown in Figure 1). This growth is translating into clinical practice as accessibility to 3D printers increases. One of the drivers for the growth 3D printing within healthcare is a trend towards development of 'personalised' medicine. Personalised medicine is "a move away from a 'one size fits all' approach to the treatment and care of patients with a particular condition, to one which uses new approaches to better manage patients' health and target therapies to achieve the best outcomes" [1]. 3D printing has been shown to be useful for: patient education [2, 3, 4] education for healthcare professionals [5], procedure planning [6, 7] and prosthesis / implant production [8] and is set to be promising in the areas of regenerative medicine and tissue engineering. Before we describe each above-mentioned section, we will highlight the workflow from medical images acquisition to application (see Figure 2).

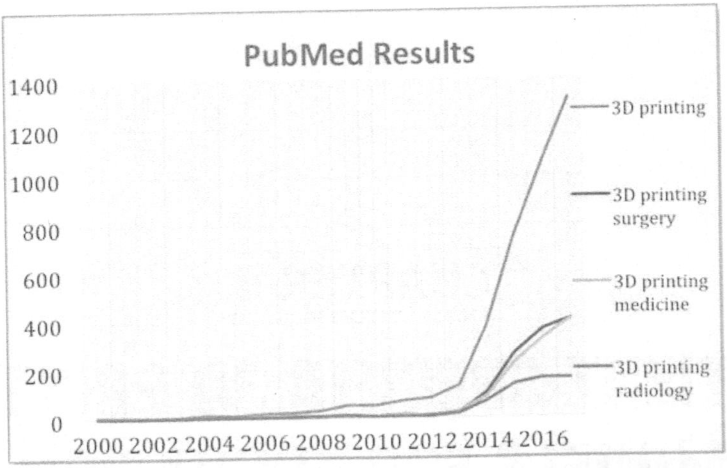

Figure 1. Chart demonstrating the number of citations in PubMed (https://www.ncbi.nlm.nih.gov/pubmed/) per annum from 2000 to 2017 in PubMed using the search terms '3D printing', '3D printing surgery', '3D printing medicine', '3D printing radiology'.

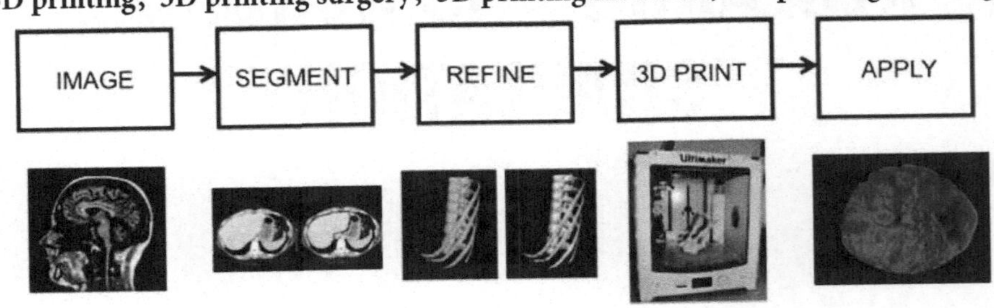

Figure 2. Outline of the workflow from medical images acquisition to application of 3D printed models.

2. WORK FLOW

The use of products derived from 3D mesh models and computer aided design (CAD) techniques in healthcare is rapidly growing. Applications include: planning surgical procedures for hepatic & renal cancer resection; innovative cardiac and vascular device testing for paediatric and adult populations; visualisation of complex head and neck anatomy for neurosurgeons; practicing procedures ex vivo; training models and educating clinicians and patients [9, 10, 11, 12, 13]. Models of heart [2], renal collecting system [14], kidney [15] and brain [16, 17] have been previously developed. Model production requires knowledge of how to segment the region of interest from medical image data, manipulate the resulting 3D model and prepare stereolithographic (.stl) files for 3D printing.

In this section we present a pipeline that converts medical images of body structures to 3D print models. Particularly, we discuss how to load and manipulate

3D medical image data, use simple processing tools to extract volumes and structures from the images, export those volumes into 3D printing software where they can refine and repair their models. We demonstrate our streamlined processing pipeline on 3D printed model of a lung, which was fabricated using filament deposition-modelling additive printing technique. This model was segmented from medical data using the freely available segmentation software Slicer.

This section will be of interest to students and professionals from medical biomedical and engineering backgrounds that wish to learn basic image processing and volume extraction techniques. The materials will make it possible to develop 3D models from medical images, which can be used as a learning aid to help visualise anatomy. As shown in Figure 2 the process starts with a 3D medical image, from which a structure will be extracted. The particular nature of the image will inform how it is processed.

2.1. Imaging

The nature of the imaging data depends on the specific imaging technology and the region of interest being imaged (see Table 1). Image resolution can vary between 0.1 and 8 mm, while image intensity can be due to density, light absorption of acoustic impedance. The main medical data file types are DICOM, NIFTI and MINC. DICOM is a universal image format and file sharing protocol, suitable for multiple image modalities and very widely used. It is easy to import into most software. NIFTI is a format designed specifically to store neuroimaging data. This format is compatible/ viewable with several specific software packages. MINC is a format used with certain brain imaging software.

Table 1. Main imaging modalities, CT-computed tomography, MRI-magnetic resonance imaging, SPECT/ PET-single photon emission computed tomography/positron emission tomography.

Modality	"Type"	Intensity from	Resolution
CT	Structural	Photoelectric coefficient = Density	0.5 - 1.5 mm
MRI	Structural/ Functional	Proton Relaxation Time	0.3 - 1.0 mm
SPECT/PET	Structural/ Functional	Particle emission from radioisotope	3.0 - 8.0 mm
Ultrasound	Structural	Reflections from changes in acoustic impedance	0.3 - 0.5 mm
Photoacoustic	Structural	Absorption of laser light by chromophores	- 0.1 mm (depth dependent)

2.2. Segmentation

The next stage of the image-processing pipeline is segmentation, which refers to the extraction of a specific 3D volume from a set of image data/slices. It is used to locate objects and boundaries in each slice that corresponds to the tissue of interest. As it is done slice by slice, a volumetric data is gradually built up. It can be used to create patient specific, highly accurate models of organs, tissue and pathology. Many software packages are available [10, 18], here we mention only Slicer. The volume can be extracted using basic or advanced segmentation techniques.

2.2.1. Basic techniques

☆ **Manual segmentation**

User identifies boundaries and manually draws around the shapes using a paintbrush tool.

☆ **Thresholding**

Pixels are partitioned depending on grayscale value. This effectively converts a grey-scale image to a binary image with one intensity representing tissue to be included in the model and the other representing that which should be excluded. This is most effective when the target tissue is a different intensity to the background.

☆ **Cropping**

Restricting the segmentation to a certain volume of space (Figure 3).

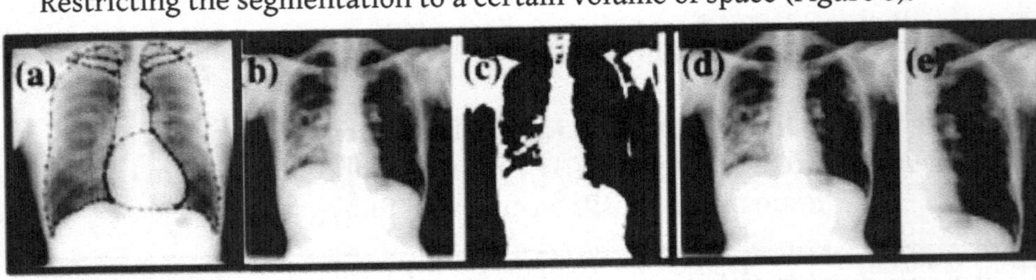

Figure 3. Examples of basic segmentation tools: (a and b) manual, (c) thresholding, (d and e) cropping.

2.2.2. Advanced techniques

☆ **Edge based methods—region growing**

'Seeds' are positioned by the user and grow to fill regions defined by boundaries in the image. Works well when regions are well defined for example contrast enhanced medium to large arteries. If the data is noisy or edges are not clear, the segmentation may 'leak'.

- **Parametric models—snakes**

The algorithm attempts to model the edges by minimising an energy term. This is minimised when the contour is on the object boundary and when the contour is as regular and as smooth as possible. It is useful for interpreting incomplete images and is robust to noise, but it can be slow to compute.

☆ **Expectation maximisation (EM)**

The algorithm finds the maximum likelihood of label distribution in a probabilistic manner. This framework is highly complex but can be a powerful tool for modelling the data accurately (Figure 4).

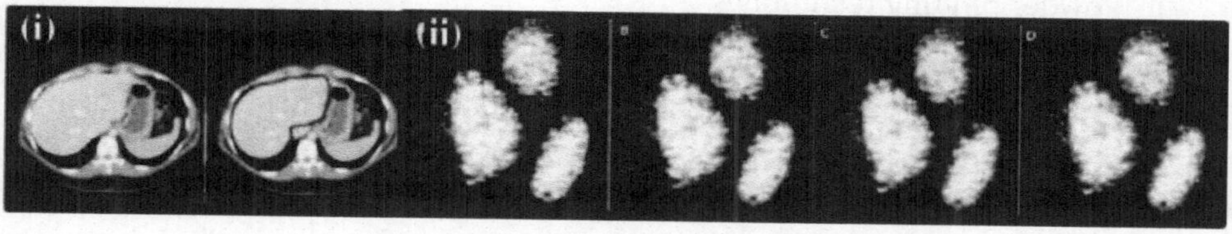

Figure 4. Advance segmentation tools:
(i) region growing, (ii) parametric models—snakes.

2.3. Refinement

Following segmentation of the 3D volume the next stage is refinement, which refers to the wide range of techniques used to convert a rough 3D segmentation into a finished, printable model. The full range of possible refinement techniques is too large to be described effectively, but key methods include:

☆ **Repairing**

Errors and discontinuities can sometimes arise in the segmentation & exporting process, which need to be repaired before printing

☆ **Smoothing**

Staircasing errors resulting from the resolution of the original medical image can be mitigated by smoothing the surface of the mesh model

☆ **Appending**

The segmentation will often only be one component of a final model. To convert the model into a final, useable form it is often necessary to combine it with other structures to, or subtracting volumes from, the mesh.

Mesh refinement can be performed using a variety of freely available software, including FreeCAD [19], MeshLAB [20], and Blender [21]. In the worked example below, extensive use is made of the Meshmixer [22], which is an easy to use mesh viewing and manipulation software, with several essential mesh refinement tools.

2.4. 3D printing

There are many different printing techniques, with many more different synonyms. It is beyond the scope of this chapter to give a complete description of every printing technique, therefore in this section we present a broad overview of current technologies. Almost all 3D printing technologies can be categorised into one of three main groups:

- ☆ Extrusion printing
- ☆ Photopolymerisation
- ☆ Powder binding techniques

The first group of 3D printers extrude a material via a print head nozzle. The material is molten and deposited on the layer underneath, where it hardens again. The most commonly used materials are thermoplastics (polylactic acid (PLA), acrylonitrile butadiene styrene (ABS)), which are deposited with a technique called "Filament Deposition Modelling". Other techniques of note are "Wire and Arc Additive Manufacturing" (used for industrial scale metal prints), as well as "Material Jetting" (which utilises inkjet print heads). Using these techniques, a multitude of materials can be printed, including metal alloys, chocolate, and even wood or ceramic composites. The second group of 3D printers selectively solidifies photopolymers. These are liquid materials that harden by exposure to light, typically ultraviolet light delivered via a laser. There are two key technologies: "Stereolithography" and "Poly Jetting". As the name of this group implies, these techniques can only print plastics. Another important technique is "digital light processing", which is very similar to Stereolithography, except that it uses a different kind of illumination.

The third group of 3D printers binds granules of the material by gluing or melting them. This method offers the widest choice of materials: glass, ceramics, many metals, and plastics. The technologies associated with this group are "Binder Jetting" and "Laser Sintering/Melting". More information about 3D printing techniques can be found on Ref. [18].

In order to print any model the file format (typically .stl or .obj) should be transformed into a language that the printer can understand (typically .gcode format). Freely available slicing software such as "Ultimaker Cura" or "Preform" help to perform this step. It converts the geometry of the model into a long series of coordinates, which the printer interprets to control the movement of an extruder or laser heads. Finally, the support material settings, print speed, temperature and other parameters should be optimised before starting to 3D print. To better understand this process, a worked example of the development of a lung model is described in the next section.

2.4.1. Working example—Lung

- ☆ Obtain DICOM imaging data (For example, from the Osirix website [23]).
- ☆ Run 3D slicer software (Download and install [24]).
- ☆ Load the DICOM data into the scene (DCM > Import > Select data > Load)
- ☆ The default setting is to have four views of the data. You can scroll through the slices using the sliders above each view. For segmentation, it's easiest to just see one view, so select the grey squares icon on the taskbar and set the view to Green slice only
- ☆ Scroll through the slices using the slider at the top of the view. We are going to be segmenting the lung on the left hand side of the view (this is actually the right lung). Find the slice at which the lung first becomes visible (A: 50.371 mm in the top right corner)
- ☆ Open the segmentation editor (Modules > Segmentation > Editor) and select OK to choose the standard colour scheme for the label map.
- ☆ Select the Level Tracing Effect Tool in the 'Edit Selected Label Map' toolbox.
- ☆ Use the tool to hover over the slice until the region of interest is outlined and left click to label the region.
- ☆ Use the slider or the scroll wheel on your mouse to go to the next slice (A: 50.957 mm) and repeat the process – hover over the region of interest until it is outlined correctly, then left click to label this region.
- ☆ Repeat on each slice in the volume where the lung is visible (up to A: 211.824 mm), highlighting and labelling the lung in each slice. Allow some time for this stage. Ensure that this is accurate – the segmentation technique has a tendency to leak or to not highlight the entire region. Use the EraseLabel and PaintEffect tools to correct errors of this form by removing leaks and filling in holes.
- ☆ Convert the label map on each slice into a 3D volume using the ModelMaker tool. (Modules > Surface Models > Model Maker)
- ☆ In the Model Maker module, ensure that the Input Volume is set to have the same name as your label. Change the Model Name to 'lung' or another suitable name, then select Apply to run the module.
- ☆ Export the model as a stereolithography file (.stl). Click File then Save. Your model should be listed as a .vtk file (lung.vtk). Ensure that the box next to this file is ticked, the other files on the list do not need to be ticked at this point. Change the file format in the drop down menu to .stl, and choose a

suitable directory where the model can be saved. Click Save to export the data.

- ☆ Open Meshmixer, free Mesh refinement software [22].
- ☆ Import your model (File > Import > Select your model from the directory > Open). You should now be able to see your lung model. There will be some errors, which we can fix in this refinement stage such as holes, non-manifold surfaces, rough edges etc. Basic commands: middle and left button on the mouse to translate, Alt and left click to rotate, scroll using middle button on mouse to zoom in and out.
- ☆ There are a number of filters that can be used to improve the quality of the model. Firstly, repair any holes using the inspector tool. (Analysis > Inspector > Auto repair all)
- ☆ Then use the RobustSmooth tool (Sculpt > Brushes > RobustSmooth). You can set the strength, size and depth of the brush according to the application. It is best to start with a low strength and a larger size, then increase the strength and reduce the size as the structure becomes smoother. Move the brush over the surface of the model in a continuous way, not spending too much time on a particular area. You should be able to see that the surface becomes visually smoother as you do this. Make sure that you save multiple versions so that is possible to go back a step if you are not happy with the result at any stage.
- ☆ The Flatten and Inflate brushes are also useful if there are unphysical holes that need to be filled in – use inflate to fill the holes then flatten and smooth so that the surface is continuous.
- ☆ Once you are happy with the model, export the volume as a .stl file for final processing and printing. (File > Export > Save)

3D printed model of a lung was fabricated by an extruded thermo-plastic polymer printer (Ultimaker2; Ultimaker, Chorley, England) using PLA filament material (PolyMax; Polymakr, Changshu, China). The Ultimaker printer is a fused deposition modelling (FDM) printer, which works by depositing layers of print material from a nozzle, which moves in the horizontal plane, onto a print bed, which moves vertically (Figures 5, 6, 7, 8, 9, 10, 11, 12). The Polymax material was chosen due to its relatively low cost, availability, and predicted ultrasonic reflectiveness. The enhanced PLA allows higher build quality and a reduced print failure rate. The STL file for a lung model was loaded into the Ultimaker CURA 3D printing software. This software allows selection of print options and generates the gcode files used by the Ultimaker printer. Due to the dimensional constraints of the Ultimaker printer, the model was printed at 90% size compared to the original CT image. The completed lung print is shown in Figure 13.

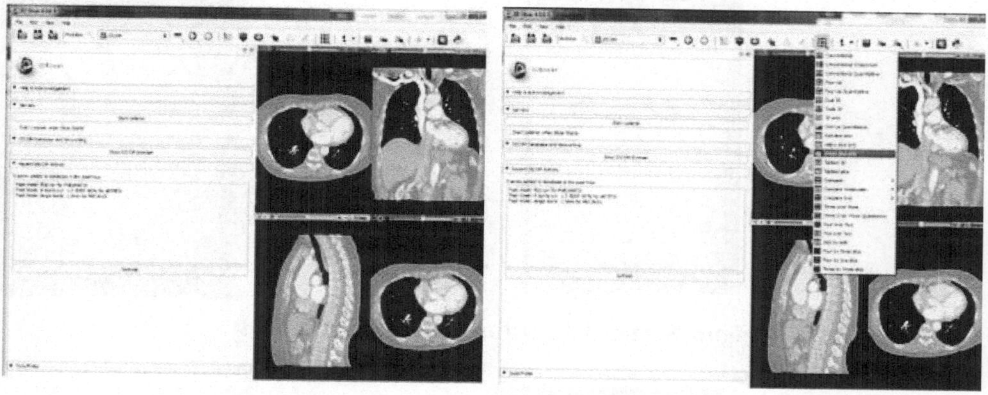

Figure 5. Steps 3–4: Importing the imaging data.

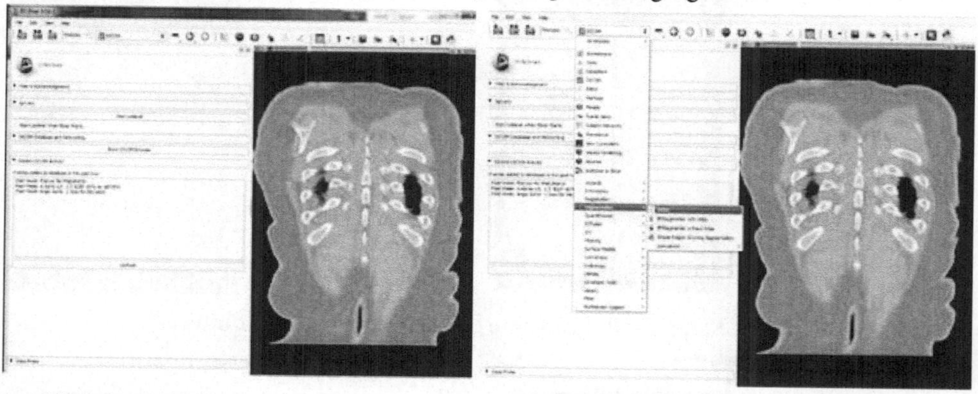

Figure 6. Steps 5–6: Finding the correct slice and open the segmentation editor.

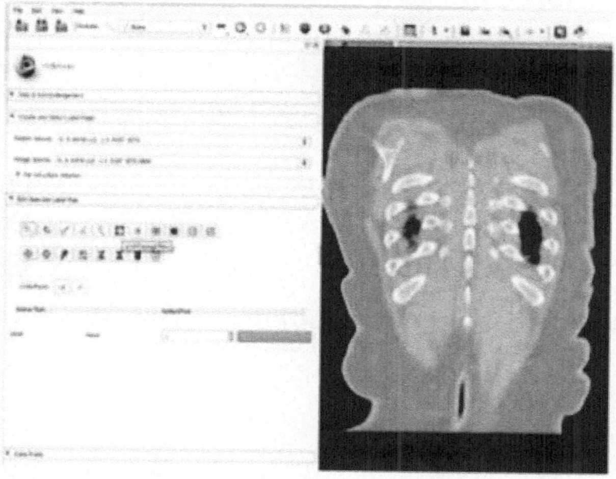

Figure 7. Step 7: Selecting the level tracing effect tool.

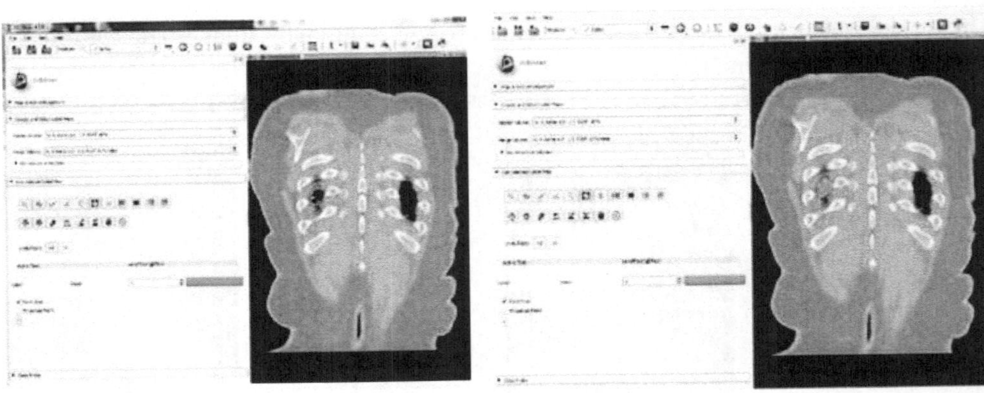

Figure 8. Step 8: Labelling the region of interest.

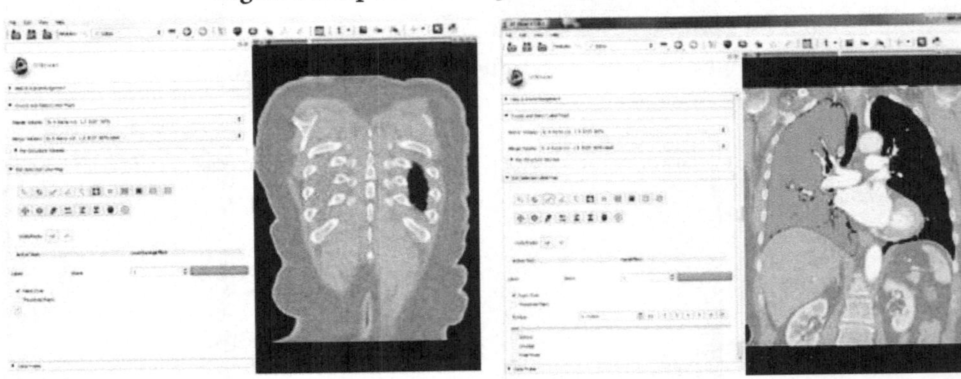

Figure 9. Steps 9–10: Continue labelling the region of interest.

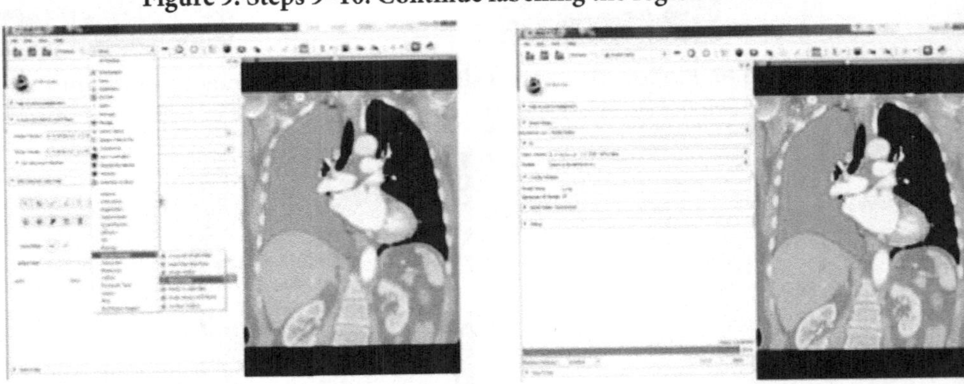

Figure 10. Steps 11–12: Creating a 3D volume using Model Maker tool.

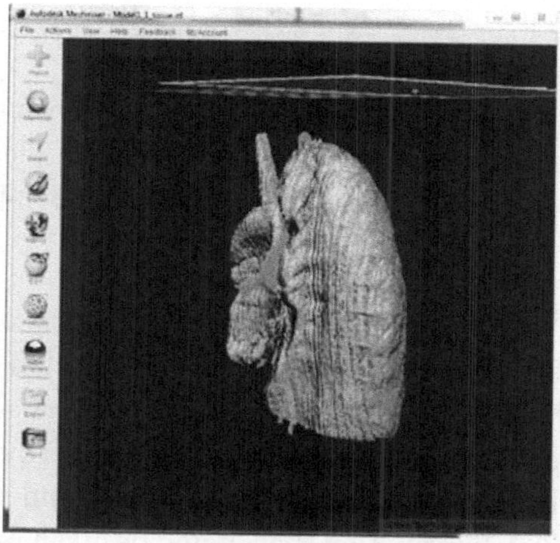

Figure 11. Step 15: Importing the model to Meshmixer software.

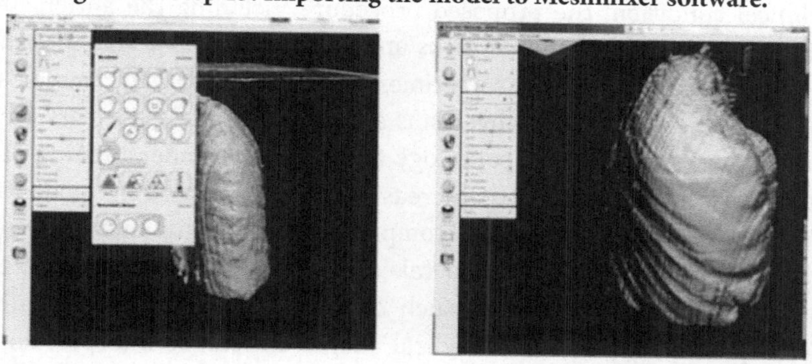

Figure 12. Steps 16–17: Repair holes and smoothing, step 19 exporting the smoothed model as a .stl file.

Figure 13. Final 3D printed model of a lung painted with acrylic paint.

3. APPLICATIONS

The following sections will describe in details the following applications: Patient Education (Section 8), Healthcare Professional Education (Section 3), Intervention Planning (Section 4), Other Applications: Implants (Section 5.1) and (Tissue Engineering) (Section 5.2).

3.1. Patient education

Guidance from both the American Medical Association [25] and the General Medical Council in the UK [26] strongly advocates a collaborative approach by physicians with their patients. It is vital that patients are provided with the relevant knowledge allowing them to engage fully in their care and to give their informed consent to treatment [27]. Information that should be given includes an explanation of the clinical condition, the proposed procedure including the anticipated post-procedural course and its benefits, risks and alternatives [27]. This information is usually communicated verbally, sometimes with the aid of diagrams or showing patients' their 'scan'. However, physicians and surgeons undergo years of training in normal human anatomy and pathology to develop an understanding of disease processes. Diseases are assessed by increasingly complex imaging modalities such as multi-phase contrast enhanced computed tomography and multisequence magnetic resonance imaging. It also takes many years to understand the vast amount of information presented in such 2D images and then conceptualise them in 3D. Consequently, patients find medical images difficult to interpret and do not enhance understanding [2]. Although many patients access additional information about their condition on the internet, this information is also often of poor quality [28].

3D printed models are proving a useful aid enhancing patient understanding of their disease. In complex diseases, replicas of the area of interest which patients can see and manipulate are thought to help understanding of the relative locations of anatomical structures, the specific areas of abnormality and the degree to which they are abnormal and what a proposed treatment (e.g. surgery) would entail [2]. A study of more than 100 parents of children with congenital heart disease in which patient-specific 3D printed models of the disease were produced and used during outpatient consultations found that 3D printed models can enhance engagement with parents and improve communication between cardiologists and parents, potentially impacting on parent and patient psychological adjustment following treatment [29]. A similar study has shown statistically significant improvements in confidence, knowledge and satisfaction amongst adolescents after consultations in which the main features of their congenital heart disease was presented using a 3D printed heart model based on their medical imaging data [29]. Similar data exists for other organ systems too. For example, a study of 7 children and 14 parents found that 3D printed patient specific livers models significantly improved parental understanding of basic liver anatomy and physiology, tumour characteristics, the planned surgical procedure, and surgical risks [30].

3.2. Healthcare professional education

3D printing is useful for education of healthcare professionals from undergraduate to expert level. It has been shown that complex anatomy is better understood from physical 3D models than 2D images. 3D printed models of segmental liver anatomy are superior to 3D virtual and 2D images for teaching anatomy and preparing for surgery [4]. The days of the traditional method of teaching surgeons and other interventional physicians known as "See One, Do One, Teach One" are gone. This method is no longer ethical or applicable mainly because of concerns for patient safety [31]. It has been replaced with competency-based training. Part of this change has been adoption of simulation-based education [32]. 3D printing can provide high fidelity and realistic models for simulation of procedures. There are numerous reports of 3D printing for simulation. For example, a recent study used box trainers and 3D-printed stomachs to assess medical students, general surgery residents, and adult and paediatric general surgeons performing a laparoscopic pyloromyotomy surgical procedure for hypertrophic pyloric stenosis, a common neonatal condition. There was a significant improvement in Global Operative Assessment of Laparoscopic Skills and Task Specific Assessments. Users felt the model accurately simulated a laparoscopic pyloromyotomy and would be a useful tool for beginners [33].

3.3. Intervention planning

Use of 3D printing is being explored for planning radiological and surgical intervention in many body systems. For example, a systematic review of 3D printed kidney models found an excellent demonstration of 3D relationships between renal tumours and adjacent anatomical structures and encouraging findings with regards to the role in surgical planning [34] Similarly, in the liver a systematic review of 3D printing has found that models have served as valuable tools in preoperative planning of surgical or interventional procedures for treatment of malignant hepatic tumours [35]. However, there are few quantitative studies and further studies with inclusion of more cases are needed [34]. 3D-printed spine models have been shown to be useful in preparing for complex spinal surgery. Using 3D-printed spine models for preparation has been reported to allow successfully performance of complex en bloc resections of primary cervical tumours [36]. In addition to open surgical procedures, 3D printing has been reported useful for planning minimally invasive and particularly endovascular interventional radiology procedures. For example; 3D-printed aortic have been used for design, planning, and/or optimization of fenestrated stent grafts [37], intracranial arteriovenous models have been found to be beneficial for radiosurgery treatment planning [38] and 3D printed models have been used to plan embolization aneurysms with challenging anatomy in the splenic artery [39] and arteriovenous malformations in the brain [40].

The most common medical application of 3-D printing is surgical guides—patient specific templates used intraoperatively to guide drilling or cutting. Using a guide specific to the patient has been shown to systematically reduce the operation time, as well as improved clinical outcomes in orthopaedic and maxillofacial surgery [41]. Another increasingly common methodology is the presurgical contouring or shaping of implants using 3D printed anatomical representations, as opposed to during the surgery itself [42, 43]. This is of particular interest in maxillofacial surgery, where a number of studies have shown reduced surgery time and improved surgical outcomes [44, 45, 46]. Similarly, bespoke 3D printed tools have been applied post surgically, through the design of bespoke external fixations for tibial fracture reduction [47].

3.4. Implants

3D printed objects can also be directly implanted into the patient to further take advantage of the ability to create bespoke, precise models. Patient specific implants (PSI) and have seen increasing interest in recent years, with numerous implants receiving FDA clearance in the first half of 2018 alone, in the wake of publication of the FDA guidance on additive manufacturing [48]. Biocompatibility is one of the major challenges of these implants, and has the devices are often made from materials which have received clinical approval previously, namely titanium and polyether ether ketone (PEEK) [41].

Using custom implants are widely accepted in maxillofacial and dental reconstruction surgery [49], in part due to the complexity of the bone and soft tissue reconstructions required. Titanium meshes have been used to create support patches to aid the repair of significant skeletal lesions [50], and splints for mandibular reconstructions [51, 52, 53, 54]. Bespoke implants have helped reduce post-operative cosmetic deformities, which are commonly associated with these surgeries [55].

The most common PSI are those created for cranioplasty to restore cranial anatomy either after surgery or repair cranial defects, as opposed to the standard treatment of autologous bone. Implants constructed out of titanium, PEEK and polymethylmethacrylate (PMMA) have all been successfully implemented surgically, and the process is becoming common practice in a number of centres [56, 57, 58]. Overall a review of custom cranial implants found the all were found to accurate and reduce operating room time, with the overwhelming majority demonstrated improvement in clinical outcomes, arising from the improved anatomical verisimilitude [41].

Neurosurgery also has the potential to benefit from 3D printing due to complexity of the anatomical considerations, with meticulous planning required due to the associated risks. Therefore a reduction of surgery time would be a considerable benefit in these cases [59]. Xu et al. [60] fabricated a 3D titanium alloy axial vertebral body that was implanted for upper cervical spinal reconstruction following a C2 Ewing sarcoma resection. A bespoke vertebral body has also been successfully implanted for reconstruction after removal of a T9 Primary bone tumour [61].

Beyond reconstructing bone and rigid structures, 3D printing methods have been developed to create bioresorbable structures, which can be used as temporary stents and splints [62, 63]. For example, a bespoke bioresorbable airway splint was successfully implanted into a child with tracheobronchomalacia [8, 64].

3.5. Tissue engineering

There are numerous applications for 3D printing technology being developed. A promising area for the integration of 3D printing technology is tissue engineering. Tissue engineering is set to provide a solution to the unmet demand for tissues and organs for regenerative medicine. This will be achieved using a combination of stem cell, bio-materials, and engineering technologies. Experts in this field believe radical improvement to tissue engineering could come from 3D printing [8]. One main problems with the synthetic scaffolds currently used is the inability to adequately mimic in vivo microarchitecture. Advances in 3D printing technology may allow production of scaffolds, which do not suffer from this problem [8].

4. LIMITATIONS

While 3D printing in healthcare is becoming more prevalent and technological advancements appear promising across a wide spectrum of applications, there are some drawbacks which must be taken into consideration. The technology is evolving and long-term evidence for the benefit of 3D printing for various applications is unknown. The potential risks of basing decisions on or carrying out procedures with poorly executed models (due to errors at any stage of the model production pipeline) is yet unknown. Another challenge is the considerable time it takes to complete the pre-print component of the pipeline. While surgery is the most complicated and expensive part of the treatment process, the increase in pre-surgical time may outweigh some of the costs saved in reducing surgery. Detailed cost effectiveness studies, which consider the increase in manufacturing capabilities and pre surgical time, as well as the reduction in operation time and improved patient outcomes, are necessary to truly evaluate the impact of 3D printing on healthcare costs.

Improving the segmentation and model design stages of the pipeline will strengthen the case for 3D printing as a cost effective healthcare technology and are therefore crucial areas of research. For example, when state-of-the-art convolutional neural networks for automatic organ segmentation are packaged for non-expert users [65] model production time may decrease. A final consideration is the range of materials available for 3D printing. Currently materials often lack the ability to mimic both the mechanical and imaging (ultrasound, optical, electrical and X-ray) properties of biological structures. Tuning the electrical or optical properties during phantom construction has been demonstrated in rigid plastic, are not readily transferrable to flexible materials. Further the choice between these properties is mutually exclusive, as the additives used control one property change the other [66, 67, 68, 69, 70]. Multimodal phantoms are an area of active research and gel wax which can be tuned to have specific optical and ultrasound imaging properties looks to be a promising material [71].

Obtaining regulatory approval has been previously outlined as a significant barrier for the widespread implementation of 3D printing technologies in medicine [72]. While these challenges are still largely in place, the publication of the FDA guidance [48] has shown a clear pathway for full regulatory approval with these devices, with over 100 devices having undergone pre-market approval.

5. CONCLUSION

3D printing is permeating nearly every aspect of medicine from education, from before treatment begins in improving education and communication, through to improving surgical planning and reducing surgery times. As the technology becomes ubiquitous, there is increased demand for extracting the relevant anatomy from medical imaging data. This places further emphasis on the tools used to automatically create representative geometry and process them in a form, which is ready to be printed. There is of course, further emphasis on demonstrating the reliability of the technologies themselves, to reduce the time taken to produce the models, and the level of expertise to use them. The review presented here gives an overview of the myriad applications of 3D printing in medicine. The workflow to create the anatomical models along with a worked example would be helpful to medical and surgical students who need access to anatomical models, and also to students from associated fields who wish to gain a hands-on understanding of surgical training and planning.

REFERENCES

1. NHS England » Personalised medicine. https://www.england.nhs.uk/healthcare-science/personalisedmedicine/.2017
2. Biglino G, Capelli C, Wray J, Schievano S, Leaver L-K, Khambadkone S, et al. 3D-manufactured patient-specific models of congenital heart defects for communication in clinical practice: Feasibility and acceptability. BMJ Open. 2015;5:e007165-e007165. DOI: 10.1136/bmjopen-2014-007165
3. Javan R, Herrin D, Tangestanipoor A. Understanding spatially complex segmental and branch anatomy using 3D printing: Liver, lung, prostate, coronary arteries, and circle of Willis. Academic Radiology. 2016;23:1183-1189. DOI: 10.1016/J.ACRA.2016.04.010
4. Kong X, Nie L, Zhang H, Wang Z, Ye Q, Tang L, Huang WLJ. Do 3D printing models improve anatomical teaching about hepatic segments to medical students? A randomized controlled study. World Journal of Surgery. 2016;40:1969-1976. DOI: 10.1007/s00268-016-3541-y
5. Nikitichev DI, Barburas A, McPherson K, Mari J-M, West SJ, Desjardins AE. Construction of 3-dimensional printed ultrasound phantoms with wall-less vessels. Journal of Ultrasound in Medicine. 2016;35:1333-1339. DOI: 10.7863/ultra.15.06012
6. Javan R, Zeman MN. A prototype educational model for hepatobiliary interventions: Unveiling the role of graphic designers in medical 3D printing. Journal of Digital Imaging. 2018;31:133-143. DOI: 10.1007/s10278-017-0012-4
7. Sheth R, Balesh ER, Zhang YS, Hirsch JA, Khademhosseini A, Oklu R. Three-dimensional printing: An enabling technology for IR. Journal of Vascular and Interventional Radiology. 2016;27:859-865. DOI: 10.1016/j.jvir.2016.02.029
8. Fishman JM, Wiles K, Lowdell MW, De Coppi P, Elliott MJ, Atala A, et al. Airway tissue engineering: An update. Expert Opinion on Biological Therapy. 2014;14:1477-1491. DOI: 10.1517/14712598.2014.938631
9. Evans LV, Dodge KL, Shah TD, Kaplan LJ, Siegel MD, Moore CL, et al. Simulation training in central venous catheter insertion: Improved performance in clinical practice. Academic Medicine. 2010;85:1462-1469. DOI: 10.1097/ACM.0b013e3181eac9a3
10. Bücking TM, Hill ER, Robertson JL, Maneas E, Plumb AA, Nikitichev DI. From medical imaging data to 3D printed anatomical models. PLoS One. 2017;12:e0178540. DOI: 10.1371/journal.pone.0178540

11. Robertson J, Hill E, Plumb A, Choong S, West S, Nikitichev D. 3D printed ultrasound phantoms for clinical training. IS&T's 32nd Int. Conf. Print. Fabr., Manchester: 2016

12. Nikitichev DI, Xia W, Daher B, Wong RYJ, David AL, Desjardins AE, et al. Placenta vasculature 3D printed imaging and teaching phantoms. IS&T's 32nd Int. Conf. Print. Fabr.; 2016

13. Rengier F, Mehndiratta A, von Tengg-Kobligk H, Zechmann CM, Unterhinninghofen R, Kauczor H-U, et al. 3D printing based on imaging data: Review of medical applications. International Journal of Computer Assisted Radiology and Surgery. 2010;5:335-341. DOI: 10.1007/s11548-010-0476-x

14. Turney BW. A new model with an anatomically accurate human renal collecting system for training in fluoroscopy-guided percutaneous nephrolithotomy access. Journal of Endourology. 2014;28:360-363. DOI: 10.1089/end.2013.0616

15. Kusaka M, Sugimoto M, Fukami N, Sasaki H, Takenaka M, Anraku T, et al. Initial experience with a tailor-made simulation and navigation program using a 3-D printer model of kidney transplantation surgery. Transplantation Proceedings. 2015;47:596-599. DOI: 10.1016/j.transproceed.2014.12.045

16. Naftulin JS, Kimchi EY, Cash SS. Streamlined, inexpensive 3D printing of the brain and skull. PLoS One. 2015;10:e0136198. DOI: 10.1371/journal.pone.0136198

17. Prados F, Nikitichev DI, Vercauteren T, Ourselin S. Patient-Specific 3D Printable Anatomical Brain Models from a Web App. Hawaii USA: ISMRM; 2017

18. 3D-MED. n.d. http://www.3d-med.uk/ [Accessed: June 8, 2018]

19. FreeCAD. n.d. http://www.freecadweb.org/ [Accessed: January 1, 2016]

20. MeshLab. n.d. http://www.meshlab.net/ [Accessed: June 8, 2018]

21. Blender. n.d. https://www.blender.org/ [Accessed: January 1, 2016]

22. Weblet Importer. n.d. http://www.meshmixer.com/ [Accessed: June 8, 2018]

23. OsiriX DICOM Viewer | DICOM Image Library n.d. http://www.osirix-viewer.com/resources/dicom-image-library/ [Accessed: June 8, 2018]

24. 3D Slicer. n.d. htttps://www.slicer.org/ [Accessed January 1, 2016]

25. Informed Consent. American Medical Association. https://www.ama-assn.org/delivering-care/informed-consent. 2018

26. General Medical Council. Consent guidance: Sharing information and discussing treatment options. n.d. http://www.gmc-uk.org/guidance/ethical_guidance/consent_guidance_sharing_info_discussing_treatment_options.asp. [Accessed: June 14, 2017]

27. Ripley BA, Tiffany D, Lehmann LS, Silverman SG. Improving the informed consent conversation: A standardized checklist that is patient centered, quality driven, and legally sound. Journal of Vascular and Interventional Radiology. 2015;26:1639-1646. DOI: 10.1016/j.jvir.2015.06.007

28. Pass JH, Patel AH, Stuart S, Barnacle AM, Patel PA. Quality and readability of online patient information regarding sclerotherapy for venous malformations. Pediatric Radiology. 2018;48:708-714. DOI: 10.1007/s00247-018-4074-3

29. Biglino G, Koniordou D, Gasparini M, Capelli C, Leaver L-K, Khambadkone S, et al. Piloting the use of patient-specific cardiac models as a novel tool to facilitate communication during clinical consultations. Pediatric Cardiology. 2017;38:813-818. DOI: 10.1007/s00246-017-1586-9

30. Yang T, Tan T, Yang J, Pan J, Hu C, Li J, et al. The impact of using three-dimensional printed liver models for patient education. The Journal of International Medical Research. 2018;46:1570-1578. DOI: 10.1177/0300060518755267

31. Kotsis SV, Chung KC. Application of see one, do one, teach one concept in surgical training. Plastic and Reconstructive Surgery. 2013;131:1194-1201. DOI: 10.1097/PRS.0b013e318287a0b3

32. Nataraja RM, Webb N, Lopez P-J. Simulation in paediatric urology and surgery. Part 1: An overview of educational theory. Journal of Pediatric Urology. 2018;14:120-124. DOI: 10.1016/j.jpurol.2017.12.021

33. Williams A, McWilliam M, Ahlin J, Davidson J, Quantz MA, Bütter A. A simulated training model for laparoscopic pyloromyotomy: Is 3D printing the way of the future? Journal of Pediatric Surgery. 2018;53:937-941. DOI: 10.1016/j.jpedsurg.2018.02.016

34. Sun Z, Liu D. A systematic review of clinical value of three-dimensional printing in renal disease. Quantitative Imaging in Medicine and Surgery. 2018;8:311-325. DOI: 10.21037/qims.2018.03.09

35. Perica ER, Sun Z. A systematic review of three-dimensional printing in liver disease. Journal of Digital Imaging. 2018. DOI: 10.1007/s10278-018-0067-x

36. Cho W, Job AV, Chen J, Baek JH. A review of current clinical applications of three-dimensional printing in spine surgery. Asian Spine Journal. 2018;12:171-177. DOI: 10.4184/asj.2018.12.1.171

37. Bastawrous S, Wake N, Levin D, Ripley B. Principles of three-dimensional printing and clinical applications within the abdomen and pelvis. Abdominal Radiology. 2018;43:2809-2822. DOI: 10.1007/s00261-018-1554-8

38. Conti A, Pontoriero A, Iatì G, Marino D, La Torre D, Vinci S, et al. 3D-printing of arteriovenous malformations for radiosurgical treatment: Pushing anatomy understanding to real boundaries. Cureus. 2016;8:e594. pp. 1-12. DOI: 10.7759/cureus.594

39. Itagaki MW. Using 3D printed models for planning and guidance during endovascular intervention: A technical advance. Diagnostic and Interventional Radiology. 2015;21:338-341. DOI: 10.5152/dir.2015.14469

40. Dong M, Chen G, Li J, Qin K, Ding X, Peng C, et al. Three-dimensional brain arteriovenous malformation models for clinical use and resident training. Medicine (Baltimore). 2018;97:e9516. pp. 1-5. DOI: 10.1097/MD.0000000000009516

41. Tack P, Victor J, Gemmel P, Annemans L. 3D-printing techniques in a medical setting: A systematic literature review. Biomedical Engineering Online. 2016;15:115. pp. 1-21. DOI: 10.1186/s12938-016-0236-4

42. Ghai S, Sharma Y, Jain N, Satpathy M, Pillai AK. Use of 3-D printing technologies in craniomaxillofacial surgery: A review. Oral and Maxillofacial Surgery. 2018;22:249-259. DOI: 10.1007/s10006-018-0704-z

43. Chana-Rodríguez F, Mañanes RP, Rojo-Manaute J, Gil P, Martínez-Gómiz JM, Vaquero-Martín J. 3D surgical printing and pre contoured plates for acetabular fractures. Injury. 2016;47:2507-2511. DOI: 1016/j.injury.2016.08.027

44. Keller EE, Baltali E, Liang X, Zhao K, Huebner M, An K-N. Temporomandibular custom hemijoint replacement prosthesis: Prospective clinical and kinematic study. Journal of Oral and Maxillofacial Surgery. 2012;70:276-288. DOI: 10.1016/j.joms.2011.06.202

45. Gil RS, Roig AM, Obispo CA, Morla A, Pagès CM, Perez JL. Surgical planning and microvascular reconstruction of the mandible with a fibular flap using computer-aided design, rapid prototype modelling, and precontoured titanium reconstruction plates: A prospective study. The British Journal of Oral & Maxillofacial Surgery. 2015;53:49-53. DOI: 10.1016/j.bjoms.2014.09.015

46. Park SW, Choi JW, Koh KS, Oh TS. Mirror-imaged rapid prototype skull model and pre-molded synthetic scaffold to achieve optimal orbital cavity reconstruction. Journal of Oral and Maxillofacial Surgery. 2015;73:1540-1553. DOI: 10.1016/j.joms.2015.03.025

47. Qiao F, Li D, Jin Z, Hao D, Liao Y, Gong S. A novel combination of computer-assisted reduction technique and three dimensional printed patient-specific external fixator for treatment of tibial fractures. International Orthopaedics. 2016;40:835-841. DOI: 10.1007/s00264-015-2943-z

48. Press Announcements - Statement by FDA Commissioner Scott Gottlieb, M.D., On FDA ushering in new era of 3D printing of medical products; provides guidance to manufacturers of medical devices; n.d.

49. Ballard DH, Trace AP, Ali S, Hodgdon T, Zygmont ME, DeBenedectis CM, et al. Clinical applications of 3D printing: Primer for radiologists. Academic Radiology. 2018;25:52-65. DOI: 10.1016/j.acra.2017.08.004

50. Horn D, Engel M, Bodem JP, Hoffmann J, Freudlsperger C. Reconstruction of a near-total nasal defect using a precontoured titanium mesh with a converse scalping flap. The Journal of Craniofacial Surgery. 2012;23:e410-e412. DOI: 10.1097/SCS.0b013e31825cef78

51. Metzger MC, Hohlweg-Majert B, Schwarz U, Teschner M, Hammer B, Schmelzeisen R. Manufacturing splints for orthognathic surgery using a three-dimensional printer. Oral Surgery, Oral Medicine, Oral Pathology, Oral Radiology, and Endodontology. 2008;105:e1-e7. DOI: 10.1016/j.tripleo.2007.07.040

52. Cohen A, Laviv A, Berman P, Nashef R, Abu-Tair J. Mandibular reconstruction using stereolithographic 3-dimensional printing modeling technology. Oral Surgery, Oral Medicine, Oral Pathology, Oral Radiology, and Endodontology. 2009;108:661-666. DOI: 10.1016/j.tripleo.2009.05.023

53. Shan X-F, Chen H-M, Liang J, Huang J-W, Cai Z-G. Surgical reconstruction of maxillary and mandibular defects using a printed titanium mesh. Journal of Oral and Maxillofacial Surgery. 2015;73:1437.e1-1437.e9. DOI: 10.1016/j.joms.2015.02.025

54. Zhou L, Shang H, He L, Bo B, Liu G, Liu Y, et al. Accurate reconstruction of discontinuous mandible using a reverse engineering/computer-aided design/rapid prototyping technique: A preliminary clinical study. Journal of Oral and Maxillofacial Surgery. 2010;68:2115-2121. DOI: 10.1016/j.joms.2009.09.033

55. Mitsouras D, Liacouras P, Imanzadeh A, Giannopoulos AA, Cai T, Kumamaru KK, et al. Medical 3D printing for the radiologist. Radiographics. 2015;35:1965-1988. DOI: 10.1148/rg.2015140320

56. Rammos CK, Cayci C, Castro-Garcia JA, Feiz-Erfan I, Lettieri SC. Patient-specific polyetheretherketone implants for repair of craniofacial defects. The Journal of Craniofacial Surgery. 2015;26:631-633. DOI: 10.1097/SCS.0000000000001413

57. Fiaschi P, Pavanello M, Imperato A, Dallolio V, Accogli A, Capra V, et al. Surgical results of cranioplasty with a polymethylmethacrylate customized cranial implant in pediatric patients: A single-center experience. Journal of Neurosurgery. Pediatrics. 2016;17:705-710. DOI: 10.3171/2015.10.PEDS15489

58. Lethaus B, Bloebaum M, Koper D, Poort-ter Laak M, Kessler P. Interval cranioplasty with patient-specific implants and autogenous bone grafts—Success and cost analysis. Journal of Cranio-Maxillo-Facial Surgery. 2014;42:1948-1951. DOI: 10.1016/j.jcms.2014.08.006

59. Mobbs RJ, Coughlan M, Thompson R, Sutterlin CE, Phan K. The utility of 3D printing for surgical planning and patient-specific implant design for complex spinal pathologies: Case report. Journal of Neurosurgery. Spine. 2017;26:513-518. DOI: 10.3171/2016.9.SPINE16371

60. Xu N, Wei F, Liu X, Jiang L, Cai H, Li Z, et al. Reconstruction of the upper cervical spine using a personalized 3D-printed vertebral body in an adolescent with Ewing sarcoma. Spine (Phila Pa 1976). 2016;41:E50-E54. DOI: 10.1097/BRS.0000000000001179

61. Choy WJ, Mobbs RJ, Wilcox B, Phan S, Phan K, Sutterlin CE. Reconstruction of thoracic spine using a personalized 3D-printed vertebral body in adolescent with T9 primary bone tumor. World Neurosurgery. 2017;105:1032.e13-1032.e17. DOI: 10.1016/j.wneu.2017.05.133

62. Ware HOT, Farsheed AC, Akar B, Duan C, Chen X, Ameer G, et al. High-speed on-demand 3D printed bioresorbable vascular scaffolds. Materials Today Chemistry. 2018;7:25-34. DOI: 10.1016/j.mtchem.2017.10.002

63. van Lith R, Baker E, Ware H, Yang J, Farsheed AC, Sun C, et al. 3D-printing strong high-resolution antioxidant bioresorbable vascular stents. Advanced Materials Technologies. 2016;1:1600138. DOI: 10.1002/admt.201600138

64. Zopf DA, Hollister SJ, Nelson ME, Ohye RG, Green GE. Bioresorbable airway splint created with a three-dimensional printer. The New England Journal of Medicine. 2013;368:2043-2045. DOI: 10.1056/NEJMc1206319

65. Wang G, Li W, Zuluaga MA, Pratt R, Patel PA, Aertsen M, et al. Interactive medical image segmentation using deep learning with image-specific fine-tuning. 2017. DOI: 10.1109/TMI.2018.2791721

66. Avery J, Aristovich K, Low B, Holder D. Reproducible 3D printed head tanks for electrical impedance tomography with realistic shape and conductivity distribution. Physiological Measurement. 2017;38:1116-1131. DOI: 10.1088/1361-6579/aa6586

67. McDermott B, McGinley B, Krukiewicz K, Divilly B, Jones M, Biggs M, et al. Stable tissue-mimicking materials and an anatomically realistic, adjustable head phantom for electrical impedance tomography. Biomedical Physics & Engineering Express. 2017;4:15003. DOI: 10.1088/2057-1976/aa922d

68. Zhang J, Yang B, Li H, Fu F, Shi X, Dong X, et al. A novel 3D-printed head phantom with anatomically realistic geometry and continuously varying skull resistivity distribution for electrical impedance tomography. Scientific Reports. 2017;7:4608. DOI: 10.1038/s41598-017-05006-8

69. Dempsey LA, Persad M, Powell S, Chitnis D, Hebden JC. Geometrically complex 3D-printed phantoms for diffuse optical imaging. Biomedical Optics Express. 2017;8:1754-1762. DOI: 10.1364/BOE.8.001754

70. Bentz BZ, Chavan AV, Lin D, Tsai EHR, Webb KJ. Fabrication and application of heterogeneous printed mouse phantoms for whole animal optical imaging. Applied Optics. 2016;55:280-287

71. Maneas E, Xia W, Ogunlade O, Fonseca M, Nikitichev DI, David AL, et al. Gel wax-based tissue-mimicking phantoms for multispectral photoacoustic imaging. Biomedical Optics Express. 2018;9:1151-1163. DOI: 10.1364/BOE.9.001151

72. Malik HH, Darwood ARJ, Shaunak S, Kulatilake P, El-Hilly AA, Mulki O, et al. Three-dimensional printing in surgery: A review of current surgical applications. The Journal of Surgical Research. 2015;199:512-522. DOI: 10.1016/j.jss.2015.06.051

CHAPTER-7

3D PRINTING OF SCAFFOLDS FOR TISSUE ENGINEERING

1. INTRODUCTION

Tissue engineering is a newly developing field of a combination of biology, materials method and engineering to develop functional substitutes for damaged tissues [1]. According to the broad range of application on cell types, it can be divided into skin, bone, vascular, kidney, and liver tissue engineering. After years of powerful progress, a set of novel tissue culture [2], replacement [3] and implantation technologies have been developed, allowing fabricating artificial extracellular matrices, namely scaffolds, to bear stem cells, growth factors, or other biological nutrients aiming at repair of tissue function. Scaffolds are bulk bioactive materials with specific porosity and structure to contribute to the formation new tissues for completing the medical task. In 2009, first artificial tissue was implanted successfully into a patient who suffered from the tracheoesophageal defect [4]. This case confirmed that artificial organs stand a chance to substitute the insufficient supply of standard organ in transplantation, which can drastically decrease the demand for living tissue. Now challenges for tissue engineering are the requirements for certain special structures, mechanical property, biocompatibility, and vascularization of tissues for implantation. In efforts to address these issues, it is important to employ an advanced manufacturing technology, which is flexible enough to build the three-dimensional (3D) structure with complex inside feature.

Reform in materials processing methods arose from the pressing needs for high-performance and multi-functional materials for broad applications in energy storage, transportation, lightweight structures, and biomedical engineering, among which 3D printing are in the highest interest by the community of material science research [5, 6, 7, 8]. In conventional processing methods, waste is cutting off from the raw material by milling, planning or grinding, and thus desired structure is obtained by these subtractive methods [9]. On the contrary, 3D printing is known as an additive manufacturing method, building the required structure layer by layer, or even pixel by pixel. The terminology "3D printing" firstly emerged was used to refer the work done at MIT in 1993, modifying a standard inkjet printer to a custom processing equipment [10]. Over last thirty years, a variety of innovative 3D printing technologies have been developed, which can be categorized into three groups including powder-based 3D printing, ink-based 3D printing, and polymerization-based printing. In all these cases, the printed structure is firstly modeled using a computer-aided design software packages, such as UG, CATIA, ProE, or other customized software. Then a ST-format file contained all the model information is exported to the 3D printing system to control the moving track of printing device and constructing the structure layer by layer.

Early use of 3D printing focused on its raid manufacture process, which is suitable for pilot production in lab or factory. Now, 3D printing is one of the most flexible technique enables direct manufacturing complex shape with high resolution, as well as processing highly customized medical products combined with image reconstitution technique. The advancement of 3D printing technologies has provided researchers and doctor's abundant tools to promote the functional scaffolds, which meet the strict criterion of tissue engineering. In addition, broadening choices in materials that can be processed by 3D printing offers researchers "recipe" to tune the biology performance of scaffolds. The ideal role of 3D printing in tissue engineering is to provide the suitable microenvironment for cells to induce cell proliferation and differentiation toward the functional tissue. There are two main modes of 3D printing using for tissue engineering currently. One is creating 3D cell-laden scaffolds that the cells are contained within the bioink. Another is fabricating molds or scaffolds, which can be cultured with cells in-vitro after fabrication [11, 12].

The main objective of this chapter is to provide a comprehensive review of the advanced 3D printing methods for tissue engineering. This chapter is structured as follows: Section 2 describes the basic need for tissue engineering. Then, a variety of advanced 3D printing methods for tissue engineering are introduced in Section 3. Finally, current issues for 3D printing methods applied in tissue engineering and potential investigations in the future are discussed.

2. KEY CONSIDERATIONS FOR TISSUE ENGINEERING

To extend the application of 3D printing into the area of tissue engineering, it is a prerequisite to have detailed knowledge of the biomaterial that is suitable for tissue engineering and can be processed by 3D printing meanwhile. The key questions to be considered for tissue engineering are components selection and mechanical features of the scaffold, which are discussed in the following sections.

2.1. Components consideration for tissue engineering

The choice of materials for tissue engineering makes up a significant portion of influence on the performance of scaffolds. Not only do the material properties should be considered, but the cellular or tissue response from the specific position should be optimizing. For all of these selected materials, nontoxicity is just the basic requirement for printing materials. In order to facilitate the cell proliferation while considering the printability from an engineering perspective, a wide range of factors should be taken into consideration when selecting printing materials for a scaffold, such as biocompatibility, bioactivity, biodegradability, and non-immunogenicity. A myriad of biomaterials suitable for scaffolds has been developed, including polymers, ceramics, metals, and even more are created each year. A range of are applied for tissue engineering.

Polymer materials have a long history in the medical industry [13]. Over last 40 years, a variety of biodegradable polymers have been developed, including synthetic and natural polymer materials. The benefits that synthetic polymers prevail over natural are that synthetic polymers can tune their initial mechanical properties and they have an abundant source of raw materials. Saturated aliphatic polyesters, such as poly (lactic acid) (PLA), polycaprolactone (PCL), poly (glycolic acid) (PGA), or their copolymers, are most frequently used tissue materials, as well as can be used as 3D printing materials [14, 15, 16]. Moreover, polymeric composites that doped with reinforcement materials, such as bioactive ceramics or carbon fibers, are allowed to be processed by 3D printing [17, 18]. The incorporation of bioactive hard phase into polymers not only enhances the mechanical property of scaffolds but also the biological performance [19].

Ceramics and bioactive glasses have been widely investigated for replacement and repair of hard tissues, such as bone tissue and teeth [20]. Traditional non-degradable bio-ceramics, such as alumina and zirconia, have high hardness and resistance to wear, making those excellent candidates in the area of joint replacement. However, their biological inertness limits the success of tissue engineering, more or less. Therefore, further efforts made by researchers were to find a ceramic with both high mechanical property and bioactivity. It is found that synthesized hydroxyapatite has close chemical components to the inorganic phase in human bone [21]. When implanted into human body, the development of the interface between HA and host tissue involves complex interactions. Solubilization of HA provides adequate beneficial ions for forming collagen and new bone tissue. Another material family used for bone regeneration is bioactive glass (45S5) whose main components are silicon dioxide and calcium oxide [22]. Both of these biocompatible ceramics and glasses have the ability to form a hydroxyl carbonate apatite (HCA) layer, which is thought to be the mechanism for their bioactive behavior.

Except for titanium and its alloys [23], which have a high bioactivity and biocompatibility to human tissue, not too much progress has been gotten for metals used in tissue engineering due to their low biocompatibility. Because of the intrinsic high strength and toughness of titanium alloys [24, 25], they are mainly used in the area of bone tissue engineering implants.

2.2. Mechanical features consideration for tissue engineering

Among the many factors need to be considered, mechanical properties of scaffolds should be tailored according to the specific site in host tissue. For example, the critical compressive strength of scaffolds used for cortical bone tissue is completely different with that for a cancellous bone tissue. For the application of segmental bone defects of cortical bone, scaffolds require compressive strength comparable to its prototype, ranging from 100 to 150 MPa along the axial direction [26, 27]. In contrast, cancellous bone has a comparatively loosen structure, which is in the range of 2.5–6.5 MPa [28]. Other mechanical properties, such as elastic stiffness, fracture toughness, and relaxation rate should also be modulated to keep consistent with original tissue [29, 30]. Because mechanical property mismatch between scaffolds and host tissue may cause stress shielding [31], which eventually results in osteoporosis.

Except for mechanical property, to achieve the goal of tissue reconstruction, scaffolds must meet some specific requirement for its architecture and internal structure. It is crucial to have interconnected pore within the bulk scaffolds transferring nutrients and oxygen for cell vascularization and proliferation. Considering the tradeoff between printing cost and biological performance ideal pore size for scaffolds ranges from 200 to 500 with a porosity between 60 and 90% [32]. However, it should be kept in mind that large pore size can facilitate cell vascularization [33]. In addition, graded channel structure can significantly promote cell migration by a capillary effect [34]. Another relevant factor is surface morphology of scaffolds, which affects the cell adhesion, can be modified plasma etching to improve its bioactivity, as well as reformed via other deposition methods [35, 36].

3. 3D PRINTING TECHNOLOGIES FOR TISSUE ENGINEERING

A range of 3D printing methods has been developed in the recent years. According to their technique characteristic, printing methods are classified into four categories, which are reviewed in the following sections, respectively.

3.1. Powder-based 3D printing

Powder-based 3D printing is characterized by using a powder bed to provide raw material, and binding powders together by polymer glue or other thermal fusion methods. It is invented in 1993 by MIT, an extra z-axis was introduced into a commercial printer by adding a height-adjustable platform, allowing printing 3D structures. In addition, the printer cartridge stored binder solution substituting original pigment. When this binder deposited on the powder bed, it can glue material together and form the desired shape. After decades of development, newer powder-based 3D printing methods, selective laser sintering (SLS), and binder jetting (BJ), are all based on this basic concept (Figure 1).

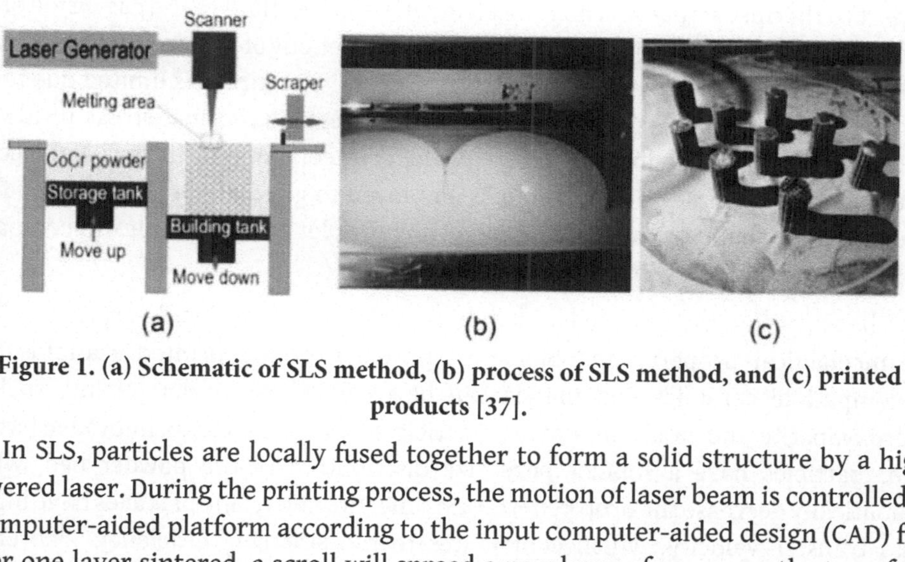

Figure 1. (a) Schematic of SLS method, (b) process of SLS method, and (c) printed products [37].

In SLS, particles are locally fused together to form a solid structure by a high-powered laser. During the printing process, the motion of laser beam is controlled by a computer-aided platform according to the input computer-aided design (CAD) file. After one layer sintered, a scroll will spread a new layer of power on the top of the previous layer, and the cycle repeats itself until the whole structure is completed. Unused particles away from heat affect zone can recycle after removing the 3D object from the powder bed, which decrease the cost of this method. Abundant processing parameter of SLS, for example, particle size, laser power, scan speed, and binder fraction, can be used to control the final structure and mechanical property of products [38]. Types of biocompatible materials that can be processed by SLS are broadening recently, from polymers and ceramics to metals. This diversity of material choice makes it possible to synthesize artificial organ matching the mechanical property of human tissue from different positions. The advantage of SLS method comes from the fact that high resolution of the laser beam. The feature size in SLS is decided both by the diameter of the laser beam and particle size, ranging from 10 to 500 μm [37, 39]. In addition, unfused powders on powder bed act as supporting materials to hold the unconnected part, decreasing minor deformation during processing. Furthermore, SLS is a one-step method that post-processing procedure, such as thermal treatment or solvent evaporation, is unnecessary when printing ceramics and metals. Polymers are the most common materials used in SLS for tissue engineering owing to its low synthesizing temperature. As for ceramics and metals, high processing heat may deteriorate the cell or drug embedded inside the printing material. For these reasons, drugs or growth factors are introduced into SLS printed scaffolds after the printing process [40].

Binder jetting is another powder-based method, which employs liquid binder to glue particles together forming the desired structure. The printer head uses either a thermal or a piezoelectric actuator to deposit binder onto the powder bed. With respect to thermal actuator, a heating element vaporizes fluid to the gas inside of the reservoir, and the increasing volume squeeze droplet out of the nozzle. The thermal method has a high-efficiency at low cost. However, its accuracy is limited due to the difficulty to control the size of the droplet, and residue thermal stress inside the binder may damage the local structure of the printed material. In the piezoelectric system, a high-accuracy piezoceramic is employed to generate pressure to the fluid reservoir. The shape and volume of the jetting droplet are more uniform compared with that in a thermal system.

Choosing suitable materials, including particle and binder, is crucial to both the mechanical property and biological property of the printed scaffolds [41]. Biocompatible ceramics and metals can be used in the binder jetting, such as hydroxyapatite and titanium dioxide. Particle size is a key factor in binder jetting. Finer particles have a smaller pore size distribution in the powder bed, which dramatically decrease the drop penetration time. However, Fine particles have higher mass transfer velocity, which contributes to the sintering efficiency. Therefore, choosing suitable particle size is a trade-off process between processing stage and thermal treatment stage. After choosing the appropriate particle materials, binder materials that used to stick particles together need to be selected as well. For the application in the medical area, the binder should not leave toxic residue when burning out, or it is nontoxic itself. Water-based binder system [42] (a water solution of an acrylic polymer) and water-soluble binder system [43] (polyvinyl acetate (PVA) or polyethylene glycol (PEG)) are two kinds of binder commonly used in ceramic casting as well as binder jetting. Binder material should have a suitable viscosity property to keep spreading from nozzles while having enough penetration ability

[44]. The shaping principle of binder jetting is more relying on physical process rather than chemical reaction, which gives rise to the flexibility in the material choice of particle used in binder jetting. Compared with SLS method, binder jetting need an extra post-processing to densify the loosen green body, because polymer binder cannot provide enough strength for the scaffolds in most cases.

3.2. Ink-based 3D printing

The ink-based method is a process that deposits fluidic materials continually or discretely out of a nozzle to a 3D platform layer by layer. It is one of the most suitable ways for processing tissue materials since it can directly print bioinks, which mixture living cells or growth factors with the liquefied material. Several 3D printing methods use this approach, including direct ink writing and fused deposition modeling (FDM).

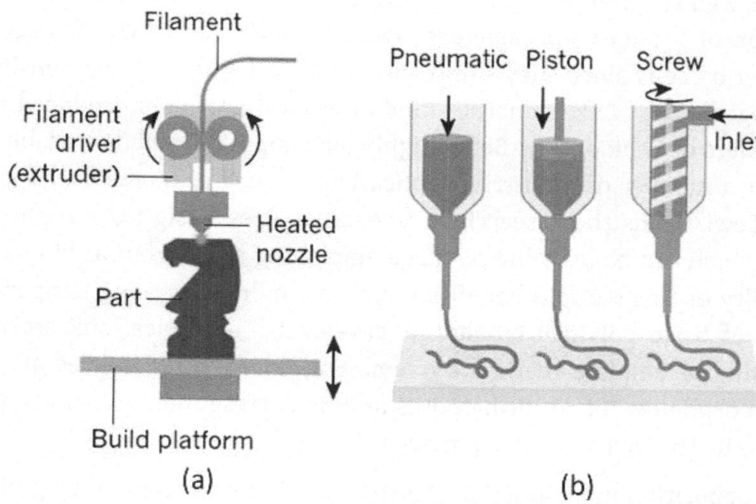

Figure 2. Ink-based 3D printing method (a) schematic of FDM method (b) schematic of DIW method [46].

Indirect ink writing (DIW) method, viscoelastic inks are squeezing out of the nozzle by the pressure from a piston, a screw, or pneumatic force as shown in Figure 2. Utilizing an easy setup, pneumatic force system has the ability to adjust the pressure in a wide range making it the most applied method in DIW. Screw system has a complicated feed module compared with other methods; however, it can provide the largest driving force that suitable for high-viscosity materials [45]. Critical to this technology is the design of the fluid property of inks. They should possess an obvious shear thinning property that allows passing through micro-size nozzle easily while recovering adequate shear strength to maintain the desired shape after inks dispensing onto the substrates. If cells are introduced into the inks as part of composition, an ideal rheology property of inks is more difficult to achieve.

Inhomogeneous cell distribution may result in an increase in viscosity locally, which causes nozzle jamming.

Cell-laden hydrogels, including chemical cross-linking [47], and molecular physical gels [48], are preferred when printing by DIW method. In bio fabrication, the selection for hydrogels is limited to the biocompatible and biodegradable. A wide range of biopolymers has been examined their viability in the medical application, such as alginate, chitosan, collagen, gelatin, and silk. Among these materials, alginate is one of the most frequently used natural biopolymers for tissue repair, wound healing and drug delivery due to its prominent biocompatibility, and the ability to differentiate cells in culture. Controllable degradation property of alginate was achieved by varying oxidation percentages of alginate hydrogel [48]. In this research, cells behavior was investigated under different concentration ranging from 1 to 20%, as well as different oxidation percentages ranging from 0 to 10%. A certain combination of these two parameters (5% of oxidation and 15% of concentration) was favored by cells since they can form a hydrogel with suitable density to hold cells homogeneously. Silk is another kind of natural polymers produced by insects such as spiders or Lepidoptera. Being highly biocompatible and degradable, silk is of interest for a number of industrial applications as well as biomedical applications. Group of Lewis leads the research in DIW area. They designed a high-resolution scaffolds, which can be used for cartilage employing silk fibroin as bioink [49]. Cell compatibility of this scaffold benefited from the mild processing temperature and avoidance of toxic polymer binders. A two-level hierarchical silk structure was created using a template method by removing micro PCL particles after printed [50]. The morphology of resulting pores and its corresponding porosity were both determined by the sacrificial PCL particles.

Fused deposition modeling (FDM) also relies on nozzle and moving platform to construct the 3D structure. However, unlike DIW, of which raw material is under liquid state, FDM need an extra heater to soften material firstly. In addition, a fan is located at the end of the nozzle, which controls the solidifying velocity of the molten material. Biocompatible polymers such as a Polylactic acid (PLA), PCL, and poly (lactic-co-glycolic acid) (PLGA) are most frequently used materials in FDM method. PCL has been widely used in dental devices and wound repair because it has high printability and excellent interactive ability with tissue.

To improve the biomedical performance of the polymer scaffolds printed by FDM, both coating [14], and doping [52] were developed for scaffolds. A better cell proliferation was obtained after surface modification by plasma treating since the improved surface roughness can adhere more cells [53]. In addition, mechanical and bioactivity properties of biopolymers are tunable by doping biocompatible reinforced particles such as HA and β-tricalcium phosphate (β-TCP) [54]. Moreover, the high concentration of reinforced phase is beneficial to the cell growth and differentiation. Generally, it is hard to attain desired mechanical property and biomedical property simultaneously. Recently, multi-material printer is developed to overcome this limitation by using a dual-printer in a single construct, as displayed in Figure 3 [51, 55]. Through this multi-material printing system, optimized carrier materials that embedded different nutrients and cell types are dispensed on discrete location in the 3D structure in one step.

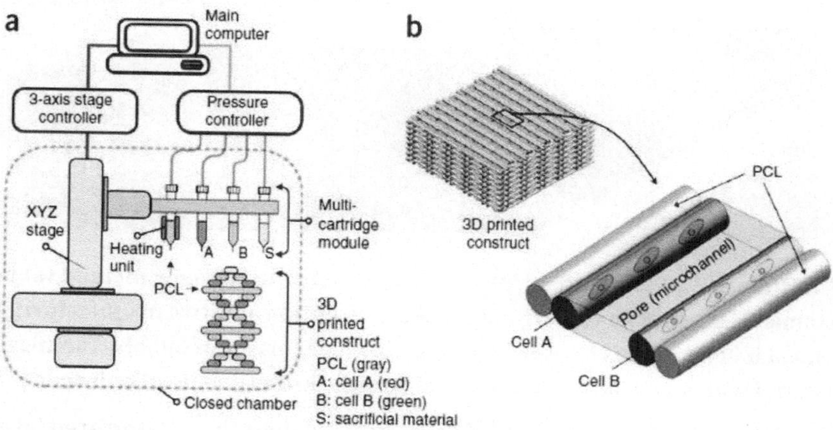

Figure 3. (a) schematic of multi-material DIW method (b) multi-material scaffold [51].

However, limitation of FDM lies in the poor choice of printing materials. Only thermoplastic materials can be fabricated by this method. Moreover, the high-temperature, ranging from 120 to 300°C, is not suitable for embedding cells or drugs inside filament when preparing scaffolds.

3.3. Polymerization-based 3D printing

The polymerization-based method starts with a process that exposing liquid photopolymer to a laser beam, then this specific exposing area would be solidified through polymer chain reaction. After repeating this process layer by layer, the final complex 3D structure can be constructed. The earliest version of this technique is stereolithography (SLA), which utilizes a low-power UV light curing photocurable polymers. Recent decades, new techniques such as two-photon polymerization (2PP) and projection micro-stereolithography (PμSL) (Figure 4), also called digital projection lithography (DPL), are developed toward a more precise and effective direction.

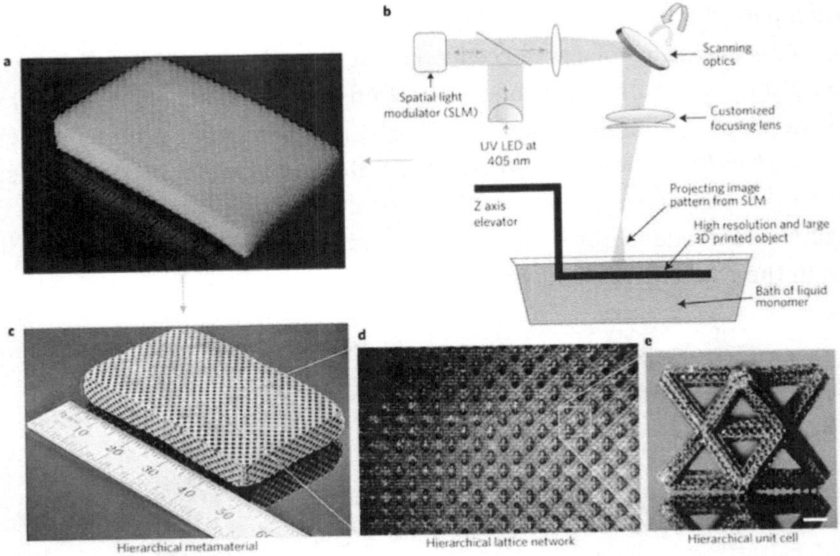

Figure 4. Hierarchical structure printed by micro-stereolithography method (a) polymer metamaterial template (b) large-area, high-resolution additive manufacturing of hierarchical metamaterials (c)–(e) optical microscope images of bulk hierarchical lattice material with a network of hierarchical stretch-dominated octet unit cells [56].

In two-photon polymerization (2PP), a long wavelength near-infrared laser beam can be focused inside of the transparent resin rather than being restricted on the surface of resin [57]. Therefore, a real 3D structure can be constructed by controlling the focal point of the laser beam. The advantage of this method is the excitation volume in 2PP is far less than other laser methods, which gives it the best resolution beyond polymerization-based 3D printing. However, the continuous processing character of 2PP confines it to be a micro-size manufacture method. Gelatin modified with methacrylamide moieties (GelMA) shows a wide range of benefits for application in tissue engineering, such as low toxicity, non-immunogenic, and tunable physicochemical properties [58], which can be used as a polymeric precursor in the 2PP method. Laura Brigo et al. successfully processed scaffolds with feature size at submicron level [59] targeting biological use. They synthesized a highly effective reaction initiator, benzylidene cycloketone-based two-photon initiator (P2CK), providing a wide processing window for photon excitation. Larger post-deformation was observed in the woodpile structure synthesized by lower laser power, which derived from the low crosslinking degree. This loose structure property is more suitable for human BJ (hBJ) foreskin fibroblasts accommodation since these cells are easily penetrating into its bulk structure.

PμSL utilizes a digital micromirror device (DMD) [60] substituting the physical masks used in lithography [61] or liquid crystal mask used in liquid crystal display

(LCD) [62] method. The basic theory used in PμSL is similar with SLA and 2PP, but the dynamic mask generator can manipulate millions of pixels at the same time rather than just one focus point, which endows PμSL the ability to process a high-resolution, large-scalability material within several minutes. A real 3D extracellular matrix (ECM) was built by DMD method to assess the difference between two-dimensional (2D) and 3D cell culture system [63]. In this research, poly (ethylene glycol) diacrylate (PEGDA), a commonly used biomaterial, was selected as lithography material to synthesize microwell-array structure. The opening space of microwell is changing along the z-position, from 250 μm at the top to 160 μm at the middle. This exquisite structure design with varying feature size was believed to have the potential to manipulate cell proliferation and cell–cell interactions. Similarly, in the 2PP method, GelMA is also a popular biomaterial employed in PμSL method [64]. Considering the different optical source in these two methods, the selection of chemicals for hydrogel preparation, such as a photoinitiator, was changing from P2CK to Irgacure 2959.

Natural structural materials, as in the case of man bone and tooth, are generally lightweight and possess a balanced combination of strength and toughness. However, synthesized bone graft materials for wound repair are relatively brittle and thus cannot match the performance of the natural part [65]. To address this challenge, a spectacular meta-structure with high tensile elasticity (>50%) was built by Xiaoyu Zheng et al. using the PμSL method [56]. This metamaterial has seven level of the hierarchy, ranging from 10 to 50 nm, and thus the mechanical property of it can match the natural materials. The high elasticity getting from the graded structure gives us the foresight to improve the mechanical property, especially the crack resistance of the synthesized biomaterials applied in the bone graft.

3.4. Four-dimensional (4D) printing

Four-dimensional (4D printing) is a recently appeared terminology in 2013 [66] and immediately attracts wide attention in different areas. 4D printing adds a new dimension, time, to ordinary 3D printed products, which allows materials responding to suitable stimuli or self-transform after possessing. It is not a totally new technique but derives from shape-morphing systems [67, 68, 69] and relies on the original 3D printing techniques. The definition of 4D printing is still in a controversy that whether the structure degradable effect can be classified into 4D process [70]. In this context, the degradation of printed material will not be discussed as 4D printing. Transformation code of 4D printed materials is hidden in the exquisite design of its structure and constituents. It offers great potential for customized medical devices given that the dynamic mechanical property of printed material accords with the behavior of living tissues [71]. In addition, the time-dependent property of 4D printing makes it suitable for long-term application embedded in human body.

One efficacious application of 4D printing is for the self-folding system [69, 70]. Two or more different kinds of materials with diverse response to outside stimuli are incorporated into an integrated structure by dual-head printers. Under the same external stimuli, the deformation difference aroused from each component will cause the structure bending or swelling toward the designed direction. This method is especially useful in cell-laden scaffolds [68]. First, a 2D thin microplate with flexible hinge was built by chemical vapor deposition (CVD) together with lithography, as presented in Figure 5. After that, cells were cultured on the thin parylene plate and thus cell traction force drove the plates folding automatically. As the lattice scaffolds can hold the cells firmly by its closed microstructure, issues with respect to how to adhere cells onto scaffolds can be avoided by this method.

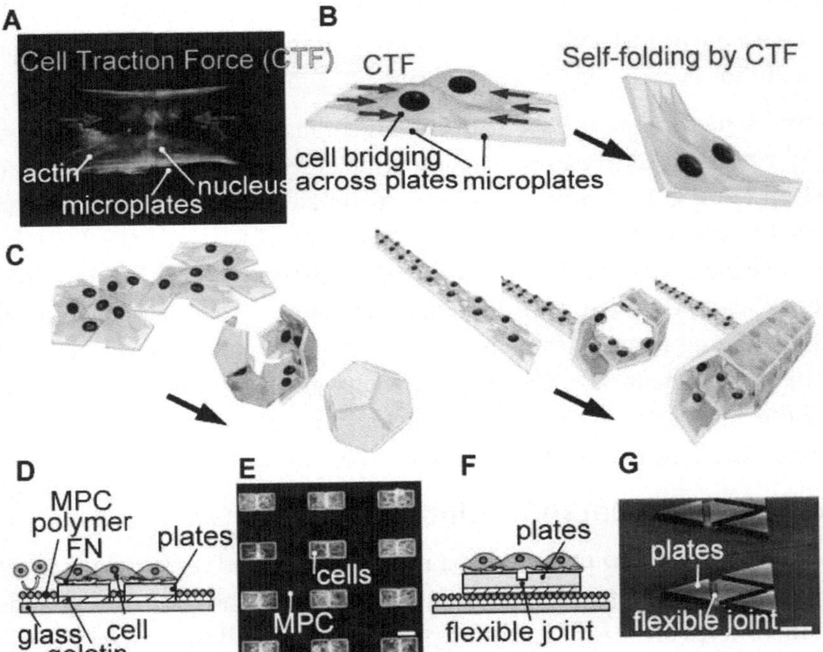

Figure 5. 4D printing for self-folding cell-laden scaffolds (a) the cells adhere and stretch across two microplates (b) the cells are cultured on micro-fabricated parylene microplates (c) various 3D cell-laden microstructures (d) schematic of the parylene microplates without a flexible joint (e) a fluorescent image merged with phase contrast image of NIH/3T3 cells patterned only on the microplates (f) schematic of the parylene microplates with a flexible joint to achieve precise 3D configurations after folding (g) a SEM image of the microplates with the flexible joint [68].

Another successful application of 4D printing in tissue engineering is making tracheobronchial scaffolds for patients who suffered from tracheobronchomalacia (TBM) [72, 73, 74]. The processing procedure including three parts. Firstly, a digital

3D model of tracheobronchial tree of patients was constructed by image software using the MRI scan data. Then the patient-specific scaffold was processed by one of the previously introduced 3D printing technologies according to the constructed 3D model. After implanted, this airway splint expanded automatically under the thermal stimuli from the internal warm organ, which leaves growing space for Malacia airway.

4. CONCLUSION

In conclusion, it is clear from the results discussed in this review that there is a huge potential for applying 3D printing in tissue engineering. 3D printing offers unique advantages toward flexible manufacturing, which can be employed to fabricate scaffolds with complex shape and internal porous structure. To improve the biological performance of printed scaffolds, it is crucial to choose suitable biomaterials introduced in Section 2, and it is equally important to select an appropriate printing technology discussed in Section 3. Although we have got great progress in the processing technique, we are still a long way from printing functional artificial tissue to completely substitute human tissue. To the best of our knowledge, 3D printing cannot build a bulk scaffold over one centimeter while possessing feature size at nanoscale. The precise control of scaffold structure, surface morphology and pore size is still a huge challenge for current 3D printing methods. In addition, post-processing is inevitable for most 3D printing methods, which limit the development of in-situ printing method. Moreover, there is a need for a significant amount of research to be carried out in order to understand the bioactive reaction between host tissue and biomaterials. With increasing research efforts in this field, we believe that future developments of novel biomaterials and processing techniques will lead us to a biocompatible artificial tissue that is smart enough to detect an event and respond to it.

REFERENCES

1. Langer R, Vacanti JP. Tissue engineering. Science. 1993;260(5110):920
2. Wei X-M, Liu Y, Wang X-Y, Gao Z-T, Yao S-M, Han J-P. Progress on research of tissue culture of Bletilla striata. Chinese Herbal Medicines. 2018;10(1):23-26
3. Fischenich KM, Lewis JT, Bailey TS, Haut Donahue TL. Mechanical viability of a thermoplastic elastomer hydrogel as a soft tissue replacement material. Journal of the Mechanical Behavior of Biomedical Materials. 2018;79:341-347
4. Mertsching H, Schanz J, Steger V, Schandar M, Schenk M, Hansmann J, Dally I, Friedel G, Walles T. Generation and transplantation of an autologous vascularized bioartificial human tissue. Transplantation. 2009;88(2):203-210

5. Kotz F, Arnold K, Bauer W, Schild D, Keller N, Sachsenheimer K, Nargang TM, Richter C, Helmer D, Rapp BE. Three-dimensional printing of transparent fused silica glass. Nature. 2017;544(7650):337-339

6. Wehner M, Truby RL, Fitzgerald DJ, Mosadegh B, Whitesides GM, Lewis JA, Wood RJ. An integrated design and fabrication strategy for entirely soft, autonomous robots. Nature. 2016;536(7617):451-455

7. Dumas M, Terriault P, Brailovski V. Modelling and characterization of a porosity graded lattice structure for additively manufactured biomaterials. Materials & Design. 2017;121:383-392

8. Raney JR, Lewis JA. Printing mesoscale architectures. MRS Bulletin. 2015;40(11):943-950

9. Chavoshi SZ, Luo X. Hybrid micro-machining processes: A review. Precision Engi-neering. 2015;41:1-23

10. Do A-V, Smith R, Acri TM, Geary SM, Salem AK. 9-3D Printing Technologies for 3D Scaffold Engineering, Functional 3D Tissue Engineering Scaffolds. Cambridge: Woodhead Publishing; 2018. pp. 203-234

11. Knowlton S, Anand S, Shah T, Tasoglu S. Bioprinting for neural tissue engineering. Trends in Neurosciences. 2018;41(1):31-46

12. Kolesky DB, Truby RL, Gladman AS, Busbee TA, Homan KA, Lewis JA. 3D bioprinting of vascularized, heterogeneous cell-laden tissue constructs. Advanced Materials. 2014;26(19):3124-3130

13. Ramakrishna S, Mayer J, Wintermantel E, Leong KW. Biomedical applications of polymer-composite materials: a review. Composites Science and Technology. 2001;61(9):1189-1224

14. Cools P, Declercq H, Ghobeira R, Morent R, De Geyter N. Acrylic acid plasma coatings for enhanced cell migration in PCL 3D additive manufactured scaffolds. Surface and Coatings Technology. 2018

15. Ferreira RTL, Amatte IC, Dutra TA, Bürger D. Experimental characterization and micrography of 3D printed PLA and PLA reinforced with short carbon fibers. Composites Part B: Engineering. 2017;124:88-100

16. Adhikari AR, Geranpayeh T, Chu WK, Otteson DC. Improved cellular response of ion modified poly(lactic acid-co-glycolic acid) substrates for mouse fibroblast cells. Materials Science and Engineering: C. 2016;60:151-155

17. Cataldi A, Rigotti D, Nguyen VDH, Pegoretti A. Polyvinyl alcohol reinforced with crystalline nanocellulose for 3D printing application. Materials Today Communications. 2018;15:236-244

18. Fu Y, Xu G, Chen Z, liu C, Wang D, Lao C. Multiple metals doped polymer-derived SiOC ceramics for 3D printing. Ceramics International. 2018

19. Elsayed H, Sinico M, Secco M, Zorzi F, Colombo P, Bernardo E. B-doped hardystonite bioceramics from preceramic polymers and fillers: Synthesis and application to foams and 3D-printed scaffolds. Journal of the European Ceramic Society. 2017;37(4):1757-1767

20. Miguez-Pacheco V, Hench LL, Boccaccini AR. Bioactive glasses beyond bone and teeth: Emerging applications in contact with soft tissues. Acta Biomaterialia. 2015;13:1-15

21. Sammons R. 3–Biological Responses to Hydroxyapatite A2–Mucalo, Michael, Hydro-xyapatite (Hap) for Biomedical Applications. Cambridge: Woodhead Publishing; 2015. pp. 53-83

22. Hench LL. The story of Bioglass®. Journal of Materials Science: Materials in Medicine. 2006;17(11):967-978

23. Pei X, Zhang B, Fan Y, Zhu X, Sun Y, Wang Q, Zhang X, Zhou C. Bionic mechanical design of titanium bone tissue implants and 3D printing manufacture. Materials Letters. 2017;208:133-137

24. Barui S, Chatterjee S, Mandal S, Kumar A, Basu B. Microstructure and compression properties of 3D powder printed Ti-6Al-4V scaffolds with designed porosity: Experimental and computational analysis. Materials Science and Engineering: C. 2017;70:812-823

25. Li Y, Yang C, Zhao H, Qu S, Li X, Li Y. New developments of Ti-based alloys for biomedical applications. Materials. 2014;7(3):1709

26. Roohani-Esfahani SI, Newman P, Zreiqat H. Design and fabrication of 3D printed scaffolds with a mechanical strength comparable to cortical bone to repair large bone defects. Scientific Reports. 2016;6:19468

27. Pilia M, Guda T, Appleford M. Development of composite scaffolds for load-bearing segmental bone defects. BioMed Research International. 2013;2013:458253

28. Giesen EBW, Ding M, Dalstra M, van Eijden TMGJ. Mechanical properties of cancellous bone in the human mandibular condyle are anisotropic. Journal of Biomechanics. 2001;34(6):799-803

29. Chaudhuri O, Gu L, Klumpers D, Darnell M, Bencherif SA, Weaver JC, Huebsch N, Lee HP, Lippens E, Duda GN, Mooney DJ. Hydrogels with tunable stress relaxation regulate stem cell fate and activity. Nature Materials. 2016;15(3):326-334

30. Das RK, Gocheva V, Hammink R, Zouani OF, Rowan AE. Stress-stiffening-mediated stem-cell commitment switch in soft responsive hydrogels. Nature Materials. 2016;15(3):318-325

31. Martin JR, Watts CD, Levy DL, Kim RH. Medial tibial stress shielding: A limitation of cobalt chromium tibial baseplates. The Journal of Arthroplasty. 2017;32(2):558-562

32. Thavornyutikarn B, Chantarapanich N, Sitthiseripratip K, Thouas GA, Chen Q. Bone tissue engineering scaffolding: Computer-aided scaffolding techniques. Progress in Bio-materials. 2014;3(2):61-102

33. Taniguchi N, Fujibayashi S, Takemoto M, Sasaki K, Otsuki B, Nakamura T, Matsushita T, Kokubo T, Matsuda S. Effect of pore size on bone ingrowth into porous titanium implants fabricated by additive manufacturing: An in vivo experiment. Materials Sci-ence and Engineering: C. 2016;59:690-701

34. Yang D, Zhao Z, Bai F, Wang S, Tomsia AP, Bai H. Promoting cell migration in tissue engineering scaffolds with graded channels. Advanced Healthcare Materials. 2017;6: 1700472

35. Wagener V, Boccaccini AR, Virtanen S. Protein-adsorption and Ca-phosphate formation on chitosan-bioactive glass composite coatings. Applied Surface Science. 2017;416:454-460

36. Yang Y, Zheng K, Liang R, Mainka A, Taccardi N, Roether JA, Detsch R, Goldmann W, Virtanen S, Boccaccini AR. Cu-releasing BG/PCL coating on Mg with antibacterial and anticorrosive properties for bone tissue engineering. Biomedical Materials. 2017;13(1):015001

37. Limmahakhun S, Oloyede A, Sitthiseripratip K, Xiao Y, Yan C. Stiffness and strength tailoring of cobalt chromium graded cellular structures for stress-shielding reduction. Materials and Design. 2017;114:633-641

38. dialkylaminoKolan KCR, Leu MC, Hilmas GE, Velez M. Effect of material, process parameters, and simulated body fluids on mechanical properties of 13-93 bioactive glass porous constructs made by selective laser sintering. Journal of the Mechanical Behavior of Biomedical Materials. 2012;13:14-24

39. Savalani MM, Hao L, Dickens PM, Zhang Y, Tanner KE, Harris RA. The effects and interactions of fabrication parameters on the properties of selective laser sintered hydroxyapatite polyamide composite biomaterials. Rapid Prototyping Journal. 2012;18(1):16-27

40. Shirazi SFS, Gharehkhani S, Mehrali M, Yarmand H, Metselaar HSC, Adib Kadri N, Osman NAA. A review on powder-based additive manufacturing for tissue

engineering: selective laser sintering and inkjet 3D printing. Science and Technology of Advanced Materials. 2015;16(3):033502

41. Inzana JA, Olvera D, Fuller SM, Kelly JP, Graeve OA, Schwarz EM, Kates SL, Awad HA. 3D printing of composite calcium phosphate and collagen scaffolds for bone regeneration. Biomaterials. 2014;35(13):4026-4034

42. Farzadi A, Solati-Hashjin M, Asadi-Eydivand M, Abu Osman NA. Effect of layer thickness and printing orientation on mechanical properties and dimensional accuracy of 3D printed porous samples for bone tissue engineering. PLoS One. 2014;9(9):e108252

43. Mieszala M, Hasegawa M, Guillonneau G, Bauer J, Raghavan R, Frantz C, Kraft O, Mischler S, Michler J, Philippe L. Micromechanics of amorphous metal/polymer hybrid structures with 3D cellular architectures: Size effects, buckling behavior, and energy absorption capability. Small. 2017;13:1602514

44. Hapgood KP, Litster JD, Biggs SR, Howes T. Drop penetration into porous powder beds. Journal of Colloid and Interface Science. 2002;253(2):353-366

45. Jungst T, Smolan W, Schacht K, Scheibel T, Groll J. Strategies and molecular design criteria for 3D printable hydrogels. Chemical Reviews. 2016;116(3):1496-1539

46. Truby RL, Lewis JA. Printing soft matter in three dimensions. Nature. 2016;540(7633):371-378

47. Hong S, Sycks D, Chan HF, Lin S, Lopez GP, Guilak F, Leong KW, Zhao X. 3D printing of highly stretchable and tough hydrogels into complex. Cellularized Structures, Advanced Materials. 2015;27(27):4035-4040

48. Jia J, Richards DJ, Pollard S, Tan Y, Rodriguez J, Visconti RP, Trusk TC, Yost MJ, Yao H, Markwald RR, Mei Y. Engineering alginate as bioink for bioprinting. Acta Biomaterialia. 2014;10(10):4323-4331

49. Ghosh S, Parker ST, Wang X, Kaplan DL, Lewis JA. Direct-write assembly of microperiodic silk fibroin scaffolds for tissue engineering applications. Advanced Functional Materials. 2008;18(13):1883-1889

50. Sommer MR, Schaffner M, Carnelli D, Studart AR. 3D printing of hierarchical silk fibroin structures. ACS Applied Materials & Interfaces. 2016;8(50):34677-34685

51. Kang HW, Lee SJ, Ko IK, Kengla C, Yoo JJ, Atala A. A 3D bioprinting system to produce human-scale tissue constructs with structural integrity. Nature Biotechnology. 2016;34(3):312-319

52. Kim MH, Yun C, Chalisserry EP, Lee YW, Kang HW, Park S-H, Jung W-K, Oh J, Nam SY. Quantitative analysis of the role of nanohydroxyapatite (nHA) on 3D-printed PCL/nHA composite scaffolds. Materials Letters. 2018;220:112-115

53. Jacobs T, Declercq H, De Geyter N, Cornelissen R, Dubruel P, Leys C, Morent R. Improved cell adhesion to flat and porous plasma-treated poly-□-caprolactone samples. Surface and Coatings Technology. 2013;232:447-455

54. Park J, Lee SJ, Jo HH, Lee JH, Kim WD, Lee JY, Park SA. Fabrication and characterization of 3D-printed bone-like □-tricalcium phosphate/polycaprolactone scaffolds for dental tissue engineering. Journal of Industrial and Engineering Chemistry. 2017;46:175-181

55. Kokkinis D, Schaffner M, Studart AR. Multimaterial magnetically assisted 3D printing of composite materials. Nature Communications. 2015;6:8643

56. Zheng X, Smith W, Jackson J, Moran B, Cui H, Chen D, Ye J, Fang N, Rodriguez N, Weisgraber T, Spadaccini CM. Multiscale metallic metamaterials. Nature Materials. 2016;15(10):1100-1106

57. Wu S, Serbin J, Gu M. Two-photon polymerisation for three-dimensional micro-fabrication. Journal of Photochemistry and Photobiology A: Chemistry. 2006;181(1):1-11

58. Yue K, Trujillo-de Santiago G, Alvarez MM, Tamayol A, Annabi N, Khademhosseini A. Synthesis, properties, and biomedical applications of gelatin methacryloyl (GelMA) hydrogels. Biomaterials. 2015;73:254-271

59. Brigo L, Urciuolo A, Giulitti S, Della Giustina G, Tromayer M, Liska R, Elvassore N, Brusatin G. 3D high-resolution two-photon crosslinked hydrogel structures for biological studies. Acta Biomaterialia. 2017;55:373-384

60. Sun C, Fang N, Wu DM, Zhang X. Projection micro-stereolithography using digital micro-mirror dynamic mask. Sensors and Actuators A: Physical. 2005;121(1):113-120

61. Mahmood MS, Celik-Butler Z, Butler DP. Design, fabrication and characterization of flexible MEMS accelerometer using multi-Level UV-LIGA. Sensors and Actuators A: Physical. 2017;263:530-541

62. Bertsch A, Jézéquel JY, André JC. Study of the spatial resolution of a new 3D microfabrication process: the microstereophotolithography using a dynamic mask-generator technique. Journal of Photochemistry and Photobiology A: Chemistry. 1997;107(1):275-281

63. Zhang AP, Qu X, Soman P, Hribar KC, Lee JW, Chen S, He S. Rapid fabrication of complex 3D extracellular microenvironments by dynamic optical projection stereolithography. Advanced Materials. 2012;24(31):4266-4270

64. Gauvin R, Chen YC, Lee JW, Soman P, Zorlutuna P, Nichol JW, Bae H, Chen S, Khademhosseini A. Microfabrication of complex porous tissue engineering scaffolds using 3D projection stereolithography. Biomaterials. 2012;33(15):3824-3834

65. Wegst UG, Bai H, Saiz E, Tomsia AP, Ritchie RO. Bioinspired structural materials. Nature Materials. 2015;14(1):23-36

66. Ge Q, Qi HJ, Dunn ML. Active materials by four-dimension printing. Applied Physics Letters. 2013;103(13):131901

67. Hu J, Meng H, Li G, Ibekwe SI. A review of stimuli-responsive polymers for smart textile applications. Smart Materials and Structures. 2012;21(5):053001

68. Kuribayashi-Shigetomi K, Onoe H, Takeuchi S. Cell origami: Self-folding of three-dimensional cell-laden microstructures driven by cell traction force. PLoS One. 2012;7(12):e51085

69. Randall CL, Gultepe E, Gracias DH. Self-folding devices and materials for biomedical applications. Trends in Biotechnology. 2012;30(3):138-146

70. Mao Y, Yu K, Isakov MS, Wu J, Dunn ML, Jerry Qi H. Sequential self-folding structures by 3D printed digital shape memory polymers. Scientific Reports. 2015;5:13616

71. Booth MJ, Schild VR, Graham AD, Olof SN, Bayley H. Light-activated communication in synthetic tissues. Science Advances. 2016;2(4):e1600056

72. Miao S, Zhu W, Castro NJ, Nowicki M, Zhou X, Cui H, Fisher JP, Zhang LG. 4D printing smart biomedical scaffolds with novel soybean oil epoxidized acrylate. Scientific Reports. 2016;6:27226

73. Zarek M, Mansour N, Shapira S, Cohn D. 4D printing of shape memory-based personalized endoluminal medical devices. Macromolecular Rapid Communications. 2017;38(2)

74. Morrison RJ, Hollister SJ, Niedner MF, Mahani MG, Park AH, Mehta DK, Ohye RG, Green GE. Mitigation of tracheobronchomalacia with 3D-printed personalized medical devices in pediatric patients. Science Translational Medicine. 2015;7(285):285ra64

CHAPTER-8

3D PRINTING PLANNING STEREOTACTIC RADIOSURGERY IN UVEAL MELANOMA PATIENTS

1. INTRODUCTION

Recent development of 3D printing enables to produce models of things, shapes, objects, and structures that before seemed more or less not possible to achieve. These technologies can build a 3D object in almost any shape imaginable as defined in a computer-aided design (CAD) file. In a basic setup, the 3D printer first follows the instructions in the CAD file to build the foundation for the object, moving the head of the printer along the x-y plane. The printer then continues to follow the instructions, moving the head of the printer along the z-axis to build the object vertically layer by layer. It is important to note that two-dimensional (2D) radiographic images, such as x-rays, magnetic resonance imaging (MRI), or computed tomography (CT) scans, can be converted to digital 3D print files, allowing the creation of customized anatomical and medical structures [1].

Charles Hull invented 3D printing, which he called "stereolithography," in the early 1980s. Stereolithography uses the ".stl" file format to interpret the data in a CAD file, allowing specific instructions to be electronically communicated to the 3D printer. Along with dimensions, the data in the ".stl" file may also contain additional information for 3D printer such as the color, texture, and thickness of the object.

Hull later founded the company 3D Systems, which developed the first 3D printer, called a "stereolithography apparatus." In 1988, 3D Systems introduced the first commercially available 3D printer, the SLA-250. Many other companies have since developed 3D printers for commercial applications. Hull's study, as well as advances made by other researchers, had revolutionized manufacturing, and it was applied in many other fields, as well as in medicine research and practice [2].

Nowadays, many of 3D printing processes are available, and all of them offer advantages and disadvantages. The model of 3D printer selected for an appliance depends on the material types that should be used, and it depends on the required layers in the finished model they are supposed to be bonded. The used 3D printing technologies within medical applications are: selective laser sintering (SLS), thermal inkjet (TIJ) printing, and fused deposition modeling (FDM).

1.1. Selective laser sintering

An SLS printing technology uses as a substrate for printing new objects, a powdered material. The laser renders the shape of the object into a single layer of powder bonding the powder particles together. Then, a new layer of powder is laid down and the process repeats on and on, building layer by layer to form the object. Selective laser sintering can be used to create metal, plastic, and ceramic objects. The accuracy of the laser and the fineness of the powder material are limited by the degree of detail. This detail enables to create especially detailed and delicate structures, like the eye globe and its structures (lens, optic disc).

1.2. Thermal inkjet printing

Thermal inkjet printing is a "noncontact" technique that uses piezoelectric, thermal, or electromagnetic technology to deposit tiny droplets of ink or other materials that are used onto a substrate according to digital instructions. Heating of the head of the printer creates small air bubbles that collapse and create pressure pulses that eject ink droplets from nozzles in volumes from 10 to 150 picoliters. Droplet size can vary by adjusting the applied temperature gradient, pulse frequency, and viscosity of the ink.

1.3. Fused deposition modeling

FDM printers are less expensive and more common than the SLS type. An FDM printer uses a head of the printer that is similar with an inkjet printer. However, instead of ink, as the printer's head moves, beads of heated plastic are released and build the object in thin layers. This process is often repeated, allowing precise control of the location and amount of each deposit to shape each layer. Since the plastic is heated as it is extruded, it bonds or fuses with the layers below. As each layer of material cools, it hardens and creates the solid object as the layers build. Depending on the complexity and cost of an FDM printer, it may have features such as multiple heads of the printer. FDM printers can use a variety of plastics [3].

2. 3D PRINTING IN OPHTHALMOLOGY

3D printing was introduced to medicine at the beginning of this century, when the technology was first used to create dental implants and individualized prosthetics. Since then, the medical applications for 3D printing technology have progressed significantly. The first published reviews describe the importance, utilization and results of 3D printing when applied in the field of orthopedic surgery, cardiovascular surgery, and tissue engineering, as well as in pharmaceutical development (new dosage forms, delivery of medicines, etc.). The current medical application of 3D printing technology is classified into several groups, such as prosthetics, implantation, and precise anatomical models; production of tissues and organs; and research in pharmaceutical industry on drug discovery, design of dispensers, and drug dosage forms [4].

2.1. 3D printing for eyewear and medical devices

3D printing technology has progressed to the point when companies print custom-made eyewear with their own design and on demand. The market for customization of eyewear has been made possible by a rapid prototyping. 3D printing enables and simplifies on demand production of medical devices in plastic or metallic form, and in the future, may represent the best way to produce artificial lenses, glaucoma valves and other personalized implants.

2.2. For education and clinical practice

3D printed models are already used in institutions for medical education—due to their accurate visualization are being used to introduce surgical techniques to trainees, young doctors or students before they start to treat patients. Surgical simulations using 3D models allow students to practice in a safe environment, until they can perform the techniques at the expected level. Hypothesis that these models can shorten the learning curve, standardize training and assessment, became true. The results show that trainers using 3D printed models have done a lot to finish their tasks better and have a better learning experience than those who used only digital models or textbooks. This suggests that using 3D models enhances the understanding of anatomical structures, their collocations, and their relationships. With the advancement of 3D printing technology, the 3D print models can be made available to improve the training of young ophthalmologists in a simulated operating theater environment, thus improving the training experience.

2.3. For printing of live cells, tissues, and organs

The development of 3D bioprinting technology, including the printing of living cells, different tissues, and even organs, is now becoming an important and expanding field of medical research. From 2012, this technology has been studied in academic circles and by biotechnology corporations (e.g., Organovo Co., San Diego, CA, USA) for possible use in tissue engineering applications, where tissues and organs are created by using 3D inkjet printing technology. The technology process is based on placing living cells onto a gel medium or sugar matrix and layer-by-layer predefined 3D structures are formed. In this way, blood vessels, bones, ears and other structures can be printed. Using 3D bioprinting technology in 2014, researchers successfully implemented a 3D skull component into a patient, with no adverse effects. This new technology represents an extension of the treatment options for creating and adapting personalized implants to the patient. The use of three-dimensional bioprinting in ophthalmology is, however, still limited, but for the generation of ocular tissues (e.g., conjunctiva, sclera, and corneas), the use of 3D bioprinting technology in the future has a great potential.

2.4. For surgical planning

In the first place, the ophthalmologist must comprehend complicated anatomical structures of the eye globe and orbit and their connection with the suspected lesion. Structural relationships observed and defined between orbital structures, muscles, vessels, and nerves can be difficult to assess fully during the planning of the orbital surgery, based solely on the 2D scans obtained. The small surgical access field for eye globe (diameter 24 mm) also means that any mistake in navigating in this structures and complicated anatomy can have potentially devastating consequences for the patient—postradiation complications. Experience proves that, for both practical and educational purposes, the creation of an anatomically personalized organ models by using 3D printing technology is very useful. This technology allows a full appreciation of anatomical relationships and collocation between tumors or lesions and other complicated surrounding, but healthy structures. Advances in 3D printing technology enable the real prototyping of various anatomical structures and allow accurate representation of the patient's current state. In surgical or irradiation planning schemes in human medicine, it will be an invaluable aid to have this real 3D organ models. It will provide a better possibility and learning experience for the doctors, physicists and surgeons [4].

We introduced the 3D printing technology into a process of planning stereotactic radiosurgery (SRS) in patients as a treatment of intraocular tumor—uveal melanoma [5, 6].

3. INTRAOCULAR MELANOMA

3.1. Definition

Intraocular melanoma is a quite rare type of cancer and is a disease in which tumor cells are formed in the part of the eye globe called the uvea (iris, ciliary body and choroid). Intermediate layer of the eye globe (uvea) contains melanocytes. Process of melanogenesis leads to produce melanin (can be found also in hair and in skin).

There are some cases in which doctors have detected intraocular melanoma during a routine eye globe examination. The chance of recovery depends on factors such as the size and cell type of the tumor. In support of classification, the staging system tumor node metastasis (TNM) is used for standardization of the tumors so the care teams can summarize information about how a tumor has spread. The information about the TNM classification is combined by a process called stage grouping. For example, intraocular melanoma grade T4 (due to classification) spreads to the orbit and extraocular tissues.

Uveal melanoma is relatively rare type of cancer, but the most common and most aggressive type of intraocular tumor in adults. The incidence of intraocular tumors varies from 0.2 to 1.0. Uveal melanoma mostly occurs in middle-aged people [7, 8].

3.2. Signs and symptoms

Most people with intraocular melanoma experience no symptoms of the disease in its early stages.

As the disease progresses, the following signs and symptoms can be seen:

☆ A growing dark spot on the iris.

☆ Change in the size or shape of the pupil.

☆ Problems with vision (blurry vision or sudden loss of vision).

☆ Floaters (spots or squiggles drifting in the field of vision) or flashes of light.

☆ Visual field loss (losing part of your field of sight).

☆ Increased intraocular pressure (secondary glaucoma).

☆ Change in the way the eye globe moves within the socket.

Only in cases when there is a massive spread outside of the eye globe, there may be pain. If someone has any of the symptoms above, it is important to visit a doctor immediately so the cause can be found and treated.

3.3. Localization of uveal melanoma

The localization of uveal melanoma is defined based on the area where the tumor is found in the eye globe and optionally based on the size of the tumor. The main localizations of the intraocular melanoma include the following tumors due to anatomical localization:

- ☆ Melanoma of the iris.
- ☆ Ciliary body melanoma.
- ☆ Choroidal melanoma (Figure 1).

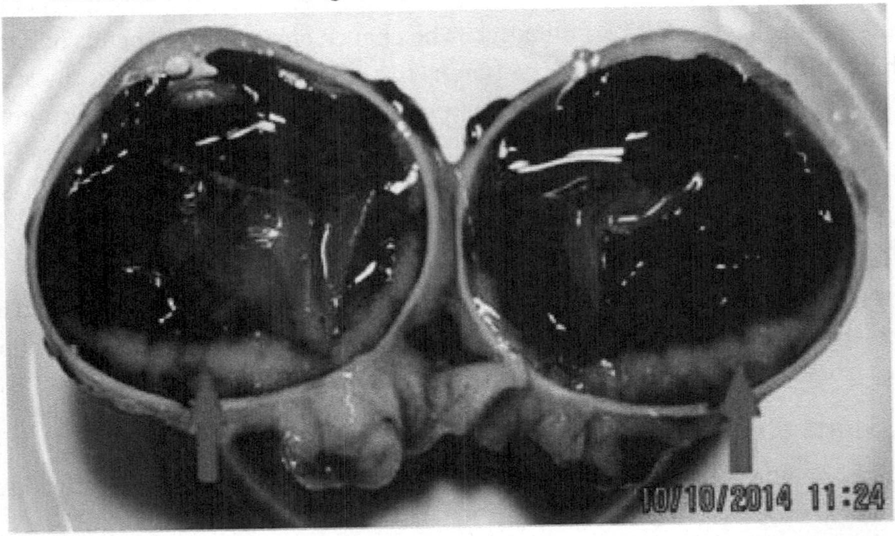

Figure 1. Enucleated eye globe with intraocular tumor—arrows shows the tumor mass (choroidal melanoma).

In certain cases, intraocular melanoma can be complicated by extraocular extension. The most frequent metastases of uveal melanoma are in the liver.

3.4. Diagnostic methods

Slit lamp examination, ophthalmoscopy and fund us photo documentation are basic examination methods.

Melanomas from choroidea can vary from dark pigmented to amelanotic, some even partially pigmented. Small choroidal melanomas are characteristic with a typical shape, where the mass under the retinal pigment epithelium is nodular, dome-shaped and well-defined. During the growth of choroidal melanoma, more irregular configurations and shapes, such as bilobular, multilobular or spongy are observed. Diffuse choroidal melanoma, whose lateral growth in choroids with minimal elevation is a characteristic feature, it is more difficult to make a diagnosis.

In many cases, the diffuse choroidal melanoma causes significant exudative retinal detachment.

If the tumor is lightly pigmented, then its abnormal vascularization can usually be detected by ophthalmoscope. Excessive choroidal melanomas usually cause changes in the epithelial pigment of the retina (e.g., drusen), atrophy patches and orange color change. These changes can occur not only in malignant but also in benign lesions. Choroidal melanoma may remain undetected under a great exudative retinal detachment or subretinal or vitreous bleeding.

A rare occurrence of advanced choroidal melanoma results in a painful blind eye globe with cataracts and proptosis, resulting from tumor transscleral orbital enlargement. In anterior choroidal melanomas, sentinel vessels (dilated episcleral vessels visible through the conjunctiva) that nourish the metabolically active tumor may occur. Transscleral growing of the anterior choroidal melanoma (predominantly via the emission channels) can be identified during the examination as a small subconjunctival area of abnormal hyperpigmentation.

3.4.1. Ultrasonography

In the case of eye globe tumors having a diameter greater than 2–3 mm, A-scan ultrasonography is suitable for diagnosis. The choroidal melanoma scan depicts a characteristic initial prominent spike, followed by low to moderate internal reflection with decreasing amplitude and significant echo. Vascular pulses can be seen as fine oscillations of the internal spiking model in the tumor area. Standard ultrasonography, currently used, has a diagnostic accuracy of more than 95%. Performing sequential A-scans with accurate dimensional measurements is a recommended complementary method after uncertain outcome of primary diagnostics.

B-scan ophthalmological ultrasonography is performed routinely for the evaluation of any suspicious mass located in the posterior segment, especially in the patients with media opacity. For the diagnosis of choroidal melanomas, this method helps to determine not only the correct diagnosis, but also to evaluate possible extraocular enlargement, to estimate the size of the tumor after periodic recurrent observations, and to appropriately schedule a therapeutic intervention.

Intraocular melanomas have several characteristics as follows:

- ☆ An acoustic quiet zone at the base of the tumor called acoustic hollowing.
- ☆ Low-to-medium reflectivity.
- ☆ Internal vascularity.
- ☆ Excavation of underlying uveal tissue.
- ☆ Shadowing of subjacent soft tissues.

Ultrasound biomicroscopy (UBM) uses high-frequency waves. This method has excellent resolution and is therefore suitable for the diagnosis of anterior ocular abnormalities. It has ability to distinguish very frontal choroidal melanomas from those that originate from the ciliary body. Moreover, it can help define the frontal plane of the tumor or help to assess angle closure glaucoma.

3.4.2. Angiography and radiography

In case of doubts, diagnostics of choroidal melanoma can be confirmed by fluorescein angiography or indocyanine green angiography, mostly in cases, when the lesions do not show pathogenic symptoms. Fluorescein angiographic changes, which can be observed in small choroidal melanomas, may be similar to some choroidal nevi. Such changes range from normal angiography findings to hypofluorescence developed secondary to blockage of the background fluorescence. In larger melanomas, a patchy pattern of early hyper- and hypofluorescence may occur, which can be followed by late intense staining. In some cases, choroidal melanomas can develop their own internal vascularization so that appear on the angiogram. Simultaneous fluorescence of the choroidal and retinal circulation in the tumor is an angiographic feature, called the "double circulation model." Its occurrence differs from choroidal melanomas.

3.4.3. Computed tomography

Computer tomography (CT) imaging method is more expensive and less sensitive compared to ultrasonography. However, it is used for scanning of the eye globe and orbit due to its ability to visualize extraocular tumor growth and to help distinguish between choroid or retinal detachment and solid tumor.

Prior to this examination, application of intravenous injection of the contrast agent is needed. As a result, contrast will cause enhancement of choroidal melanoma, whereas in case of exudation will not.

3.4.4. Magnetic resonance imaging

Magnetic resonance imaging (MRI), also used for scanning the eye globe and orbit, is even more expensive than CT scanning and less sensitive than ultrasound. However, MRI uses surface coil imaging and gadolinium as a contrast material for improving its resolution. Within the MRI, high-density lesions represent T1-stage pigmented melanomas, and low density can exhibit pigmented melanoma at T2 stage. In many cases, MRI is used to determine the extrascleral melanoma enlargement or to distinguish surrounding fluid from the tumor.

3.4.5. Biopsy and genetic analysis

Practice usually does not require the use of a fine needle biopsy and an incisional biopsy. However, these additional examination methods may be useful in differential diagnosis. They are used to distinguish amelanotic melanomas from metastatic tumors and as complementary tests, if the results of other tests are ambiguous. It is possible to achieve more than 95% accuracy in tumors larger than 3 mm using both types of biopsy. Comparing both methods, incisional biopsy is more invasive and comes with higher complications rate, but has lower rate of falsely positive or negative results. Intralesional or perilesional hemorrhage are the most common complications, a fine needle biopsy is not connected with higher risk of spreading cancer cells in the case of choroidal melanoma.

Following biopsy, the prompt treatment is indicated with an aim to prevent extrascleral extension.

Nowadays, genetic analysis and karyotyping of biopsy samples are becoming more important. Various studies have documented that Chromosome 3 monosomy in the choroidal tumor is associated with a significantly higher risk of developing metastatic process. Nowadays, sadly, there is still no effective treatment for metastatic disease available.

3.5. Treatment opportunities

The important prognostic indicators they initiate the following therapy for posterior uveal melanoma are the age and the volume (size) of the tumor. Many studies are documenting that over 50% of patients with uveal melanoma die because of direct or indirect reasons (e.g., metastasis) within 15 years after the therapy, although radical surgery (enucleation) or other therapeutical methods are used.

In turn, modern diagnostic tools, they include ophthalmological examination, CT, and MRI brought most vital advances in the ability of primary uveal melanoma diagnosis. The diagnostic methods have radically improved in the last decades, and different types of radio surgery (external beam, charged particle or brachytherapy) have become the preferred treatment in significant number of patients with uveal melanoma. One of the main reasons for the development of alternative therapies is the objective of promoting survival and maintaining vision in patients who have experienced uveal melanoma. Among other things, many types of radiation are used in today's treatment of posterior uveal melanoma. One of the representatives of the "conservative" approach is stereotactic radiosurgery (SRS) with the usage of a linear accelerator.

In addition to SRS, the gamma-knife radiosurgery method is used. Both methods provide good local control with survival rates comparable to other treatments. SRS of extracerebral lesions, such as uveal melanoma, has been invented about 20 years ago. An alternative treatment that has been proven to treat middle and large posterior choroidal melanoma is the SRS method. Another method is the plaque radiotherapy, when the eye globe salvation is achieved. It is used especially in cases of tumor location outside the optical disc or the macula. The positive consequence for the patient is the useful vision that can be retained after treatment. Linear accelerator therapy with a single fraction of the intraocular tumor is considered to be a relatively unusual approach to the treatment of uveal melanoma. For treatment planning coordinates is used an image fusion of a contrast-enhanced MRI and CT. When using a collimation system in an operation, it is important to achieve spatial accuracy when administering one fraction [9, 10, 11].

4. STEREOTACTIC RADIOSURGERY FOR INTRAOCULAR TUMORS

Stereotactic radiosurgery was initially developed in 1949 by Lars Leksell, the Swedish neurosurgeon, who treated small targets of tumors located in the brain. It was a new step in radiotherapy methods. Nowadays, the stereotactic radiosurgery (or external beam irradiation with protons or helium ions) is a regularly used option in the treatment of mainly medium-sized choroidal melanomas. In many cases, it has been applied for larger tumors. Eye globe and orbit scanning by CT or MRI is necessary for verifying of extraocular extension. It is necessary to use these techniques to differentiate between choroidal tumor and primary versus secondary retinal detachment. Every patient with a medium-sized melanoma is sent to chest X-ray, liver ultrasound, and general examination but also to PET/CT to detect possible metastases.

Methods like MRI, CT and also digital subtraction angiography (DSA) are included in imaging equipment. Structures of the eye globe and lesions are defined, visualized and localized. Software for target verification, which is used in conjunction with stereotactic frame system and imaging (CT, MRI) modalities, is used to determine the coordinates of the lesion (tumor) target into the stereotactic frame reference system.

The radiosurgical treatment is calculated by physicist by planning system for treatment (TPS) with the 3-D dose distribution. The result is superimposed onto the certain patient's anatomical status to get an appropriate radiation dose for tumor and also for risk structures (optic nerve, lens, chism, etc.) and radiosurgical procedure [12, 13, 14].

5. 3D PRINTED MODELS OF THE INTRAOCULAR MELANOMA FOR PLANNING OF STEREOTACTIC RADIOSURGERY TREATMENT

Stereotactic radiosurgery (SRS) is a technically challenging therapeutic irradiating method. It complements or supplies (replaces) classic surgical intervention. It is used for single high-therapeutic irradiation dosage to an exact specified volume, while risk organs and structures are contemporary protected. A prerequisite for this method is a special software and hardware equipment of workstation and professional experiences of specialists of different fields (neurosurgeon trained in stereotactic radiosurgery, radiation oncologist, ophthalmologist, radiologist, clinical physicist and registered nurse trained for radiosurgery).

The surgical operation involves the selection and preparation of the patient prior to surgery intervention, which consists from processing of health and imaging documentation of the patient. It is necessary to analyze the patient's disease and the patient's indication by the Indicating Commission (BTB). The Commission decides about the indication in which the members are neurosurgeons trained in radiosurgery, radiation oncologists, ophthalmologists, radiologists and clinical physicists. The progress committee decides on the basis of a recommendation on the appropriateness of an eye globe oncological surgery, the listed group of experts will evaluate the suitability of classical surgery, stereotactic radiosurgery operations, fractionated stereotactic radiosurgery, intensity modulated radiotherapy (IMRT) or three-dimensional comfort radiotherapy (3D-CRT).

Patients indicated for stereotactic radiosurgical intervention are hospitalized for inpatient care. This is a short-term hospitalization, most often lasts 3 days. The patient admission involves interview with the patient with detailed information about the course of operation, performance benefits as well as acquaintance with potential acute and late postoperative complications (adverse effects). The patient subscribes the informed consent.

Patient's admission in hospital bed department (clinical care) is carried out 2 days before the surgery. A detailed clinical examination will be done, brought documentation is studied, the missing examinations are completed and a preoperative pharmacotherapy treatment in hospital bed department is placed on. Findings, surgery and documentation will be inserted into the hospital information system. The day before the stereotactic radiosurgery (SRS) will be initiated the patient's pharmacotherapy—premedication. Within the preoperative premedication, the patient is receiving the antiedematotic therapy, which intensity depends on the size, location of the lesion and the presence of edema. The presented therapy is continued in the day of surgery and the following day.

Each patient's record includes the tumor size, tumor volume, the maximum height of the tumor, age and gender, the presence and the extent of secondary retinal detachment, and the possible signs of extrascleral extension. The very important step of SRS is tumor volume calculation in each patient directly by results of CT and MRI examination. It is the basic step involved to the stereotactic planning scheme.

Immobilization of the affected eye globe for stereotactic irradiation is achieved by mechanical fixation of the sutures to the Leibinger stereotactic frame. Vicryl sutures from four extraocular muscles (m. rect. sup., m. rect. inf, m. medialis and m. temp.) are placed through the conjunctiva and the lids. The stereotactic frame is fixed to the head, and the sutures are tied to the stereotactic frame. The day after stereotactic irradiation is the patient examined by an ophthalmologist (the slit lamp examination, ophthalmoscopy, and intraocular pressure measuring) and is released for home treatment, he gets antibiotic and corticosteroid eye drops.

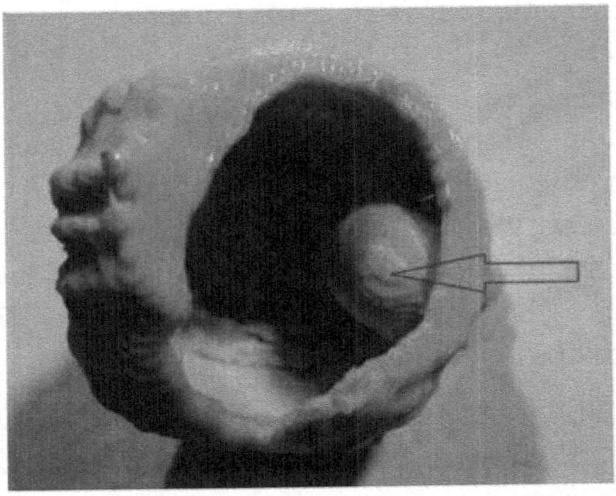

Figure 2. 3D printed model of the eye globe with intraocular tumor—arrow shows the tumor mass (melanoma).

In software for data segmentation (3D Slicer, freeware version 4.5.0+), virtual 3D models of the eye globe were created. Imported data set came from CT (computer tomography) with accuracy 1 mm scan thickness. Visualized and created model has basic anatomical structures like globe, corneal segment, lens, optic nerve and tumor mass inside the created 3D model of eye globe. Extraocular muscles and vitreous body are not included in the model. Eye globe anatomical structures—lens and optic nerve—are very important for orientation. They are necessary visible points at the small printed 3D model because the model itself has the same size as the normal human eye globe—from anterior to posterior part is 24 mm. Virtual 3D model of the eye globe is then sliced. Additional support is calculated. The next step can be model preparing for printing in 3D printing slicing software Simplify3D. After creating and refining our models, for process of 3D printing, we used fused deposition modeling (FDM) technology. The 3D models in our study are printed on ZYYX 3D Printer by

Magicfirm LLC using one extruder heated to 215–230°C. Chosen material is polylactic acid (PLA) with low deformability during rapid temperature changes and contributing high accuracy of the model. One layer of the model thickness of in vertical line is 100 μm. This provided an ideal proportion of a ratio between accuracy and velocity of a printing process. Estimated time of printing process in 3D printer is usually from 15 to 30 min per one model (Figures 2 and 3).

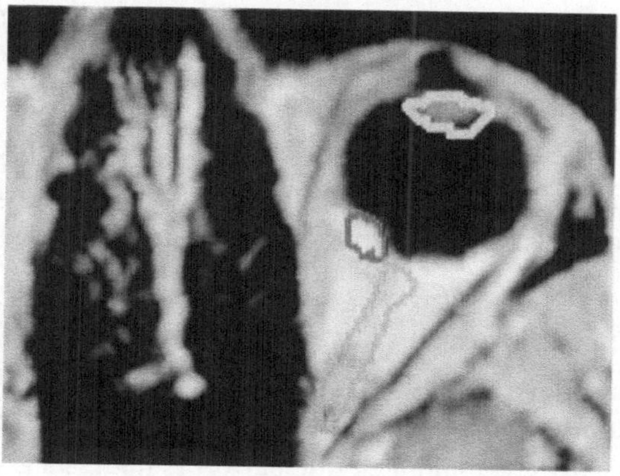

Figure 3. MRI scan of eye globe with uveal melanoma (choroidal melanoma stage T1b)—red color, lens—green color, optic nerve—yellow color.

Stereotactic one-day session therapy of intraocular uveal melanoma, based on CT and MRI images, is a precise treatment option and safe. The results of the treatment were summarized in the study and demonstrated that the local control was excellent. Among other results, they have further shown that in patients with an unfavorable tumor size and localization near critical structures such as the optic nerve, lens and macula, visual acuity has been reduced. In the patient's follow-up after 6 months, visual acuity was evaluated. The study included a group of 77 patients in whom 85.5% of patients had a visual acuity of 0.1 or higher before radiotherapy. In the cases with intraocular uveal melanoma, these patients may be recommended to undergo LINAC-based one-day session stereotactic irradiation, especially in the cases of medium-sized uveal melanomas. Important advantage of this treatment method is preserving of the eye globe [5, 6].

6. DISCUSSION

In the treatment of the uveal melanoma patients, stereotactic irradiation has been used over 20 years. Over time, the therapeutic single dose decreased and stabilized at 35.0 Gy. In the studies, this reduction did not lead to tumor control reduction. The studies also documented that hypofractionated treatment is beneficial. This treatment uses a very large fraction to stabilize the intraocular uveal melanoma cell lines.

In the last years, additional interest in fractionated stereotactic radiotherapy (SRT) was obtained. Feasible fractionation advantage is used by Linear accelerators (LINAC). Nowadays, therapeutic schemes for hypofractionated therapy for five fractions with total doses from 50.0 to 70.0 Gy are employed by most LINAC studies. Different studies report local tumor with control rates more than 90% by the efficacy of SRT in uveal melanoma treatment. In treatment of melanoma close to the optic disc, juxtapapillary choroidal melanoma, this stereotactic radiotherapy offers another non-invasive alternative to radical surgery—enucleation—or to brachytherapy treatment with high tumor control rate [9, 15]. The gamma knife radiosurgery and SRS are evaluated as the alternative in patients with large tumors (stage T3) which are too large for conventional brachytherapy. Studies confirm with their results that treatment with SRS may also be recommended in specific cases of uveal melanoma [16, 17].

Interesting is the comparison with other forms of radiotherapy. Typical side effects of radiation from SRT include: radiation retinopathy, cataract development, secondary neovascular glaucoma, and opticopathy. The result may be loss of visual acuity. In the worst cases, secondary enucleation is necessary. However, even with these risks, SRS and SRT are among the most effective treatments for uveal melanoma. The SRS method belongs to one of the new methods; therefore, it is needed to record and compare results with other methods by a multicenter studies. Treatment of uveal melanoma using a combination of stereotactic photon therapy and CT and MRI scans is considered very safe and accurate and guarantees excellent local control. A reduction in visual acuity was observed in a large number of patients. This condition was a result of the unfavorable size and position of the tumor near critical structures, such as the optic nerve and the macula. Obtained patient observations confirmed that stereotactic irradiation for uveal melanoma using LINAC is feasible and well tolerated. This treatment preserves the patient's eye globe and is offered in the case of moderate or unfavorably localized uveal melanomas [5]. Optical neuropathy may rarely occur after radiosurgical treatment of lesions that are located near visual pathways. A highly effective method in these cases for the treatment of small and moderate uveal melanomas is a one-step LINAC-based SRS with a single dose of 35.0 Gy. This treatment involves the use of a mechanical immobilization system with four stitches [18].

3D printing has gained its new role in ophthalmology by confirming and demonstrating the real shape and size of tumors by creation of accurate 3D models. With current advances of precise modeling process, material to build the 3D models can be printed at 600 μm resolution. This approach is already in use within the other fields of ophthalmological surgery. An example is the use in cataract surgery—development of Cana's Ring (CR) and a special expansion device for 3D pupils [19].

Another example of the practical use of 3D printing in ophthalmology is the eye fundus examination. The correct collocation of the patient, the smartphone and the Volk spherical lens is extremely important. To secure the correct position, a 3D printed adapter from plastic was created to connect the lens with the smartphone together [20].

Different studies evaluated the efficacy of stereotactic radiosurgery as a treatment modality for uveal melanoma. Tumor control rates 5 and 10 years after therapy were over 90% in several studies. Results of these studies did not show a significant difference between radiogenic side effects of SRS and other forms of radiotherapy. Majority of visual acuity decline and necessity of secondary enucleation are neovascular glaucoma. Radiation therapy of stereotactic photon beams for the treatment of intraocular melanoma is considered effective nowadays. However, it is still recommended for future studies and follow-up actions to focus on finding optimal treatment modalities [21]. 3D printed model of eye globe with tumorous mass is due to our experience helpful during the stereotactic planning process. Planning of stereotactic radiosurgery is based on fusion of CT and MRI data. Understanding of the collocation of all the structures of the eye globe with critical structures inside, such as lens, optic disc and the tumor size and its location, is a basic condition for the planning software. By introducing a new 3D modality for visualization of the eye globe with the tumor, we provided real model of the eye globe for the specialists, which enabled them increase effectiveness of the planning process during the stereotactic radiosurgery. Virtual 3D model can be seen on computer only, but real printed 3D model of the eye globe held by the physician and physicist in their hands gives them a better perspective of the real situation inside the eye globe, like the size, the distance of the tumor to critical structures, the lens, optic disc and optic nerve [5].

7. CONCLUSIONS

The number of patients in our study with reduced visual acuity will probably increase in the future. Created model was used for comparison with pictures of the eye globe after enucleation due to secondary complications (amotio retinae). In future, we plan to make a collection of interesting models of rare cases for purposes of education.

Current possibilities of creating the real-sized or even also enlarged 3D models of the affected eye globe help ophthalmologists and other specialists to comprehend and visualize the size and localization of the tumor, has been rather difficult before, solely with the 2D radiographic imaging. The 3D model of an eye globe with an intraocular tumor has an important role during the process of preparing a stereotactic radiosurgery planning scheme. This is a step forward for the team of specialists responsible for management and treatment of a patient with malignant melanoma. Implementation of a 3D model in the process of individualized irradiation planning of stereotactic surgery has a great potential for the future.

REFERENCES

1. Ventola CL. Medical applications for 3d printing: Current and projected uses. Pharmacology & Therapeutics. Oct 2014;39(10):704-711
2. Prasad TS, Sujatha G, Muruganandhan J, Patil S, Raj AT. Three-dimensional printing in reconstructive oral and maxillofacial surgery. The Journal of Contemporary Dental Practice. Jan 1, 2018;19(1):1-2
3. Valverde I, Gomez G, Suarez-Mejias C, Hosseinpour A-R, Hazekamp M, Roest A, et al. 3D printed cardiovascular models for surgical planning in complex congenital heart diseases. Journal of Cardiovascular Magnetic Resonance. Feb 3, 2015;17(Suppl 1):P196. Available from: https://www.ncbi.nlm.nih.gov/pmc/articles/PMC4328535/ [Accessed: Apr 14, 2018]
4. Huang W, Zhang X. 3D printing: Print the future of ophthalmology. Investigative Ophthalmology & Visual Science. Aug 1, 2014;55(8):5380-5381
5. Furdová A, Sramka M, Thurzo A, Furdová A. Early experiences of planning stereotactic radiosurgery using 3D printed models of eyes with uveal melanomas. Clinical Ophthalmology (Auckland, N.Z.). Jan 31, 2017;11:267-271
6. Furdová A, Furdová A, Thurzo A, Šramka M, Chorvát M, Králik G. Possibility of 3D printing in ophthalmology—first experiences by stereotactic radiosurgery planning scheme of intraocular tumor. Ceská a Slovenská Oftalmologie. Fall 2016;72(3):80-84
7. Shields JA, Shields CL. Intraocular Tumors: An Atlas and Textbook. Philadelphia, Pa.: Lippincott Williams & Wilkins; 2008. 598 p
8. Furdova A, Sramka M. Uveal Malignant Melanoma and Stereotactic Radiosurgery: Intraocular Uveal Melanoma and one-Day Session Stereotactic Radiosurgery at Linear Accelerator. Saarbrücken: LAP LAMBERT Academic Publishing; 2014. 188 p
9. Mueller AJ, Talies S, Schaller UC, Horstmann G, Wowra B, Kampik A. Stereotactic radiosurgery of large uveal melanomas with the gamma-knife. Ophthalmology. Jul 2000;107(7):1381-1387 (discussion 1387-1388)
10. Tokuuye K, Akine Y, Sumi M, Kagami Y, Ikeda H, Kaneko A. Fractionated stereotactic radiotherapy for choroidal melanoma. Journal of the European Society for Radiotherapy and Oncology. Apr 1997;43(1):87-91

11. Furdova A, Strmen P, Waczulikova I, Chorvath M, Sramka M, Slezak P. One-day session LINAC–based stereotactic radiosurgery of posterior uveal melanoma. European Journal of Ophthalmology. 2012;22(2):226-235

12. Zehetmayer M, Kitz K, Menapace R, Ertl A, Heinzl H, Ruhswurm I, et al. Local tumor control and morbidity after one to three fractions of stereotactic external beam irradiation for uveal melanoma. Journal of the European Society for Radiotherapy and Oncology. May 2000;55(2):135-144

13. Hungerford JL. Current trends in the treatment of ocular melanoma by radiotherapy. Clinical & Experimental Ophthalmology. Feb 2003;31(1):8-13

14. Selby BP, Sakas G, Walter S, Groch W-D, Stilla U. Pose estimation of eyes for particle beam treatment of tumors. In: Bildverarbeitung für die Medizin 2007 [Internet]. Berlin, Heidelberg: Springer; 2007. pp. 368-373. (Informatik aktuell). Available from: https://link.springer.com/chapter/10.1007/978-3-540-71091-2_74 [Accessed: Apr 14, 2018]

15. Krema H, Somani S, Sahgal A, Xu W, Heydarian M, Payne D, et al. Stereotactic radiotherapy for treatment of juxtapapillary choroidal melanoma: 3-year follow-up. The British Journal of Ophthalmology. Sep 2009;93(9):1172-1176

16. Dieckmann K, Georg D, Zehetmayer M, Bogner J, Georgopoulos M, Pötter R. LINAC based stereotactic radiotherapy of uveal melanoma: 4 years clinical experience. Radio-therapy and Oncology. May 2003;67(2):199-206

17. Furdova A, Sramka M, Chorvath M, Kralik G, Krasnik V, Krcova I, et al. Stereotactic radiosurgery in intraocular malignant melanoma--retrospective study. Neuro Endocrinology Letters. 2014;35(1):28-36

18. Furdova A, Sramka M, Chorvath M, Kralik G. Linear accelerator stereotactic radiosurgery in intraocular malignant melanoma. Austin Journal of Radiation Oncology and Cancer. 2015;1(2):1009

19. Canabrava S, Diniz-Filho A, Schor P, Fagundes DF, Lopes A, Batista WD. Production of an intraocular device using 3D printing: An innovative technology for ophthalmology. Arquivos Brasileiros de Oftalmologia. 2015 Dec;78(6):393-394

20. Myung D, Jais A, He L, Blumenkranz MS, Chang RT. 3D printed smartphone indirect lens adapter for rapid, high quality retinal imaging. Journal of Mobile Technology in Medicine. 2014;3(1):9-15

21. Zehetmayer M. Stereotactic photon beam irradiation of uveal melanoma. Developments in Ophthalmology. 2012;49:58-65

CHAPTER-9

THREE-DIMENSIONAL PRINTING AND NAVIGATION IN BONE TUMOR RESECTION

1. INTRODUCTION

One of the challenges of orthopedics and traumatology has been recreating a preoperative planning scenario that contemplates 3D space, without the need to use cadaverous specimens, and being able to reproduce it in the real world. Thanks to the advent of new technologies within the computer age, a new research model was born: "in silico" [1], which means "done by computer or via computer simulation." In addition to the phrases "in vivo" and "in vitro" of Latin, which are used in systems biology and refer to experiments done in living organisms or outside living organisms, respectively, "in silico" is translated to "in silicium," which refers to the material from which semiconductors are made, alluding to computer information storage. This creates the concept of turning a virtual computer scenario into a research lab.

One of the possibilities of in-silico research and development consists of performing virtual 3D models that faithfully represent reality. In the field of orthopedics and traumatology, this type of tool opens room for many new developments, such as a virtual 3D model based on images of computer tomography (CT) and magnetic resonance that could simulate the specific anatomical structure of a patient.

In orthopedics surgery, the precision of the cut when performing a particular osteotomy can have a great impact in the final surgical outcome. For example, in bone tumor resection, the osteotomy should leave free margins outside tumoral contamination but at the same time respect as much healthy bone tissue as possible. Therefore, the use of a simulation scenario to determine where and how to execute an osteotomy with the greatest precision possible would mean a clear advantage when planning and performing this type of orthopedic surgery.

In the same way that a global positioning system (GPS) can orientate a person through an unknown path, an intraoperative simulation scenario would be able to guide the path that the cutting saw must follow during a surgical procedure. The surgical planification can be done in a virtual 3D model and then executed under virtual navigation [2, 3, 4, 5, 6, 7].

However, virtual navigation based on images contemplates a unique point in space, therefore guiding the tip of the surgeon's instrumental through the bone surface. For this reason, it is necessary to mark the planned scheme on the patient's cortical surface and then execute it with a conventional saw under navigated guidance. This makes the level of precision and accuracy questionable, when performing a uniplanar, biplanar or multiplanar osteotomy [5, 6, 7].

Many experiments have been conducted to measure the precision associated with an osteotomy, which has been virtually planned and performed under navigation. Wong et al. were one of the first to report that planned tumor resections were facilitated with the use of intraoperative navigation and that this gives clinical benefits [8, 9]. These advantages were also probed in the computer-assisted surgery of the pelvis and sacrum [10, 11]. Postoperative computer tomography (CT) images of the patient can be superimposed to the original preoperative 3D scenario, allowing digital measuring of the distance between the target plane and executed plane [12].

A common approach to measure surgical precision is to use the resected specimen, which contains the tumor, obtained from surgery. Possible methods include the histological evaluation of the specimen, which determines the distance between the cut edge and the tumor by microscopy imaging measurement [12]. However, this does not allow comparison with the planned osteotomy, as it only reports the effective oncological margins. Another method consists of CT scanning the surgical specimen and adding it to the preoperative 3D scenario as another 3D piece [13]. The specimen piece location is manually matched against the original bone structure by an operator, obtaining the best possible image registration. The distance between the target plane and the executed plane can then be measured virtually.

The possibility of measuring the surgical precision obtained after performing an osteotomy is a key factor to allow continuous improvement in the field of orthopedic surgery. This is necessary for evaluating the effect of new instruments and surgical technologies, new surgical techniques, or any development that needs to be tested. Moreover, surgeons could experience a better learning curve for navigation systems if they could study the results they obtain.

In this context, the technological advancement provided by 3D printing represents an interesting possibility. 3D printers have become very popular and it is common to find them in clinical research environments. The concept of rapid prototyping (RP) allows to create very specific models based on computer-assisted designs (CAD). In this way, a virtual scenario can be reconstructed for visualizing the bone, the tumor and the planes of osteotomies and then print those structures, giving the surgeon the possibility to have in his hands a 3D model, in real scale, that faithfully represents the patient's situation. As described below, we can use this prototype model to simulate a surgery. The principles of navigation based on images (combining magnetic resonance images [magnetic resonance imaging MRI] with CT studies) can be applied to the RP model [6, 7, 8, 10]. In this way, a correspondence between the real structure of the RP model and its 3D reconstruction is obtained in the navigation computer. Finally, surgeons can carry out the reproduction of a virtual plan in a prototyped bone.

The main objectives of this chapter are to evaluate:

- ☆ if it is possible to create a 3D model in a virtual scenario based on CT scans and MRI, which can simulate the morphology of the bone structure, the tumor and the oncological margins, obtaining a virtual three-dimensional preoperative planning;
- ☆ if it is possible to print a 3D prototype (in this case, a model of proximal tibia) in real scale, which is representative of an oncological patient. This includes contemplating the healthy bone tissue, the tumor tissue and the oncological margins;
- ☆ if it is possible to use this prototype to test osteotomies under navigated guidance based on images.

2. MATERIALS AND METHODS

The procedure included three distinctive stages: virtual 3D planning, printing a rapid prototyping (RP) model and image-based navigation.

2.1. Virtual 3D planning

The workflow needed to obtain a 3D reconstruction of a patient's anatomical structure can be divided into three phases, which are explained below.

2.1.1. Image acquisition phase (computer tomography and magnetic resonance)

Both a conventional magnetic resonance imaging (MRI) study and a multiknee computer tomography (CT) were performed for an 11-year-old patient. The patient's only symptom was recurrent knee pain, and diagnosis of metaphyseal osteosarcoma in the proximal tibia was confirmed by biopsy. The treatment chosen was transepiphyseal tumor resection and reconstruction of the defect with structural allograft bench [14].

Toshiba CT scanner (Aquilion, Japan) was used and the tomographic acquisition protocol was the following: FOV 32 cm; pixel size 0.625 mm; KV 120; 100 mAs; thickness of cut 1 mm; and height and width of image (512 × 512 pxl). Images were digitized in digital imaging and communication in medicine (DICOM) format. Siemens resonator (Avanto, Germany) was used and the digital resonance acquisition protocol was the following: FOV: 32 cm; pixel size 0.75 mm; thickness of cut 1 mm; height and width of image (256 × 256 pxl); in time T1.

2.1.2. Image segmentation phase

Once the image files are obtained, the objective is to eliminate elements that look like bone but are not bone. This process, known as image segmentation, is done by establishing a colorimetric assessment. Figure 1 shows bone tissue represented in yellow, eliminating other elements such as cartilage, muscle, fat, skin or other elements that do not belong to the bone, such as the CT scanner lead. This procedure is performed manually by the operator and determines the final reconstruction of the bone, eliminating structures foreign to the bone tissue that can alter the anatomical form of the bone.

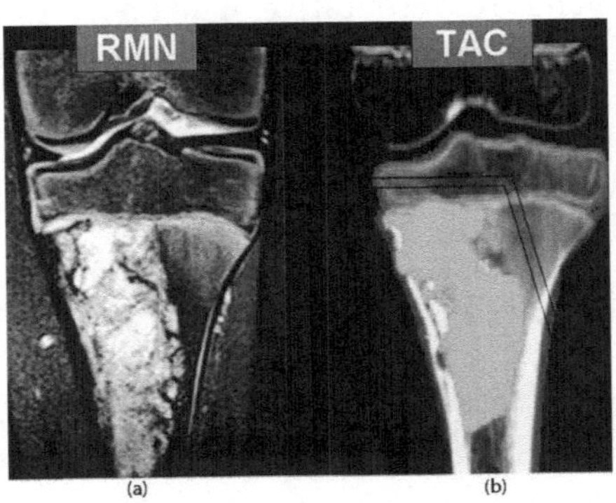

Figure 1. (a) Magnetic resonance where hyperintense tumor lesion observed in the proximal region of the tibia. (b) Tomography image. The area corresponding to the tumor was painted (segmented) in green, while healthy bone was segmented yellow. This process is repeated in each of the cuts.

The bone tissue was segmented from CT scan, while the tumor tissue was segmented from MRI. Image fusion was then performed using a mold that overlaps both images in the proper place.

2.1.3. 3D reconstruction and planning phase

Once the entire volume of 2D is segmented, this volume is transformed into a 3D bone structure within a virtual scenario. By representing all three axes of space [15], this scenario implies an advance in the way of measurement previously used with CT images and 2D MRI, obtaining a virtual 3D bone that aims to reproduce reality. In this way, the bone morphology and the tumor as a structure can be obtained in a virtual space.

2.2. Printing the RP model

The rapid prototyping model was created using Z-Printer Spectrum Z-510 printer, in 1 hour and 47 min of printing. It was printed in two halves to include within it four colors that define what the tumor is and where the oncological margins are. Two planes were created near the tumor: one in red (unsafe margin of 3 mm) and the other in blue (safe margin of 3 mm). The tumor was colored in green while the rest of the structure corresponding to healthy tissue was printed in white.

The printing technique consists mainly of spilling a film of 0.1 mm of dust ZP131 on the base of more dust (flat) of 2 cm of thickness. Next, a liquid adhesive is printed on that thin film leaving the shape of the first layer of the object under study (proximal tibia), which will correspond to a biplane section of 0.1 mm in thickness. This act is repeated consecutively until the finished piece is obtained. In this work, the tibia was printed in two halves.

Due to the fragility of the newly manufactured piece, the next step consisted of structural fixation with isocyanato, polyol and acetone. Then, both halves were bonded together with Z-Bond 101 glue, as shown in Figure 2. The precision of the printed piece was validated by measuring four-known distances in silico and comparing them in the RP model created.

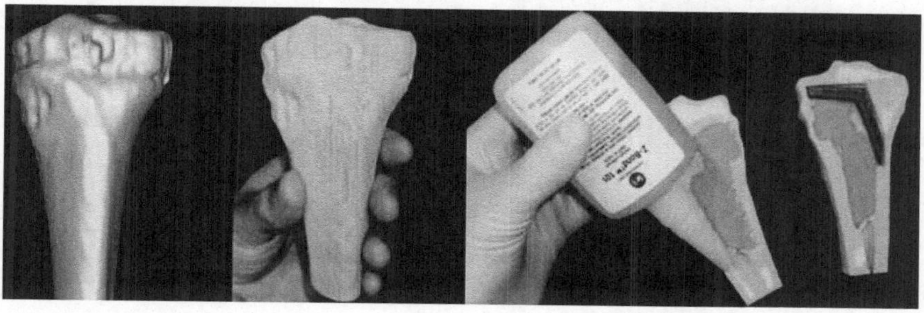

Figure 2. Three-dimensional in-silico planning model printed as RP model.

2.3. Image-based navigation

Once the RP model was prepared, it was fixed to a work table with a vice. The 3D preoperative surgical plan was loaded to Stryker Navigation System II (Figure 3). Then, an infrared surface recorder (tracker) was firmly fixed to the RP model with two pins and a label (Figure 4). This device generates a 3D coordinate system surrounding a specific area of interest, thus allowing to guide the navigator by matching the 3D preoperative scheme with the RP model.

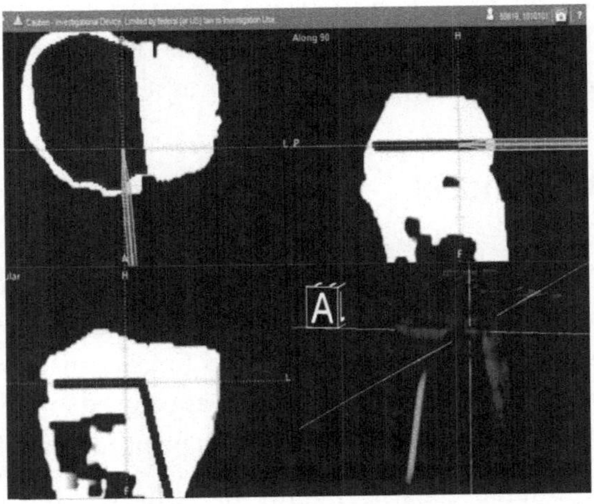

Figure 3. Real-time three-dimensional cut map displayed in the browser during cutting. The instrument colored in green corresponds to the blade of the saw. The safe margin is observed in blue.

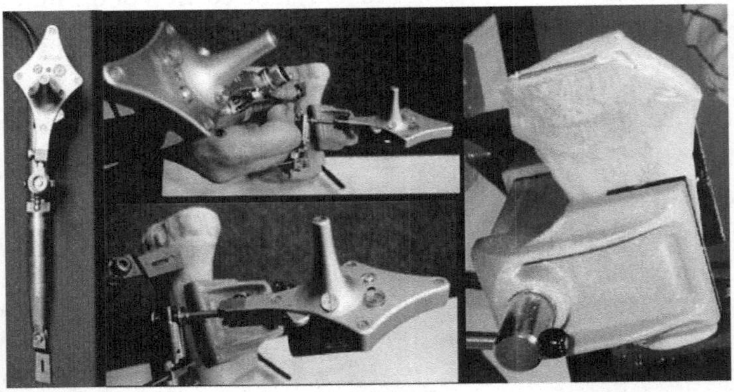

Figure 4. Saw with recorder. The RP model is osteotomized with a saw, which has a recorder attached to guide the direction of the cut in a navigated manner.

After this, the model is ready to be osteotomized with navigation tools. The same tracking procedure was carried out on a cutting saw, with a 1.5 mm thick blade. In this case, an appropriate hook was used to fix the crawler to it, allowing one to see the edge of the saw and its trajectory in real time in the screen of the navigation equipment. In this way, the operator visualizes the virtual plane of the previously planned cut from the scenario in silico and directs the saw accordingly, contemplating all three spatial axes.

Once the navigated osteotomy was performed, the piece was opened to verify the correspondence of the virtual planning and the RP model.

3. RESULTS

3.1. Virtual 3D planning

The segmentation of healthy and tumoral bone tissue was performed. In this way, a 3D piece was built in a virtual scenario, where surgical approaches were evaluated until a final system of osteotomies was chosen. The choice was done contemplating the oncological margins with colors for both distal and proximal osteotomies to the tumor.

The preoperative planning, shown in Figure 5, was designed based on this 3D scenario. Two virtual osteotomy planes 3 mm thick were created. This thickness represents the cutting saw and its oscillation. The original coronal plane corresponding to the MRI at T1 was used to determine the distance of the oncological margin. This distance determines the location of the planes. In this way, we can establish a 3D cutting plane that considers the distance between the tumor tissue and the necessary oncological margin.

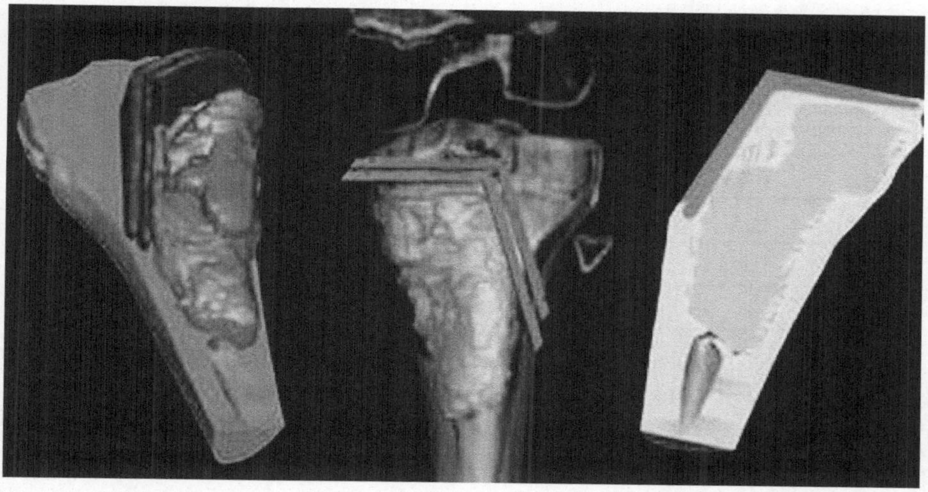

Figure 5. Three-dimensional preoperative planning model "in silico." Healthy bone tissue is represented in gray and tumoral tissue in green. The blue cut plane represents safe oncological margin, and the red cut plane represents unsafe oncological margin.

3.2. Printing the RP model

By printing the bone in two halves, we were able to create a solid structure that corresponded in size and shape to what was observed in our virtual scenario.

The validation of said procedure was carried out by comparing the measurement of four known distances in silico to their measurement in the RP model (Figure 6). These differences are stated in Table 1.

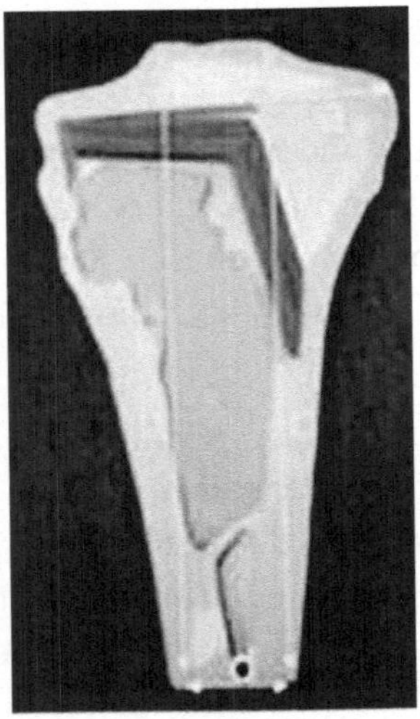

Figure 6. Validation distances.

Table 1. Measurement of distances to validate the RP model's precision.

	Virtual measurement (mm)	Difference between virtual and physical measurement (mm)	Relative error (%)
A	62.8	0.5	0.7962
B	112.2	0.1	0.0891
C	110.8	0.3	0.2708
D	19	0.5	2.6316

The average relative error, considering the virtual measurement as true measurement, is of 0.9469%. The level of error is considered tolerable for this application; therefore, it can be concluded that the prototyping method shows good reproduction of patient's structure in real scale.

3.3. Image-based navigation

The surgeons were able to reproduce the preoperative plan on the RP model by using a conventional saw under intraoperative navigation guidance. The procedure was validated by opening the piece and confirming its correlation with the previous biplanar osteotomy planned in silico. This is represented as the distance between each cut edge of the piece and its corresponding oncological margin planes.

The distal plane showed correspondence with the safe margin limit of the planned plane, not reaching the red plane (unsafe margin) at any point. The proximal plane obtained was parallel to the planned plane but showed a translation of 1 mm toward the healthy tissue side. In both cases the unsafe margin, colored in red, maintained all its integrity.

In Figure 7(a), we can see both halves of the RP model after osteotomies were performed, identifying clearly where the margin was greater than planned for the proximal plane and how the distal plane was correctly executed. In Figure 7(b), we can see the real specimen of proximal tibia presenting osteosarcoma, for comparison.

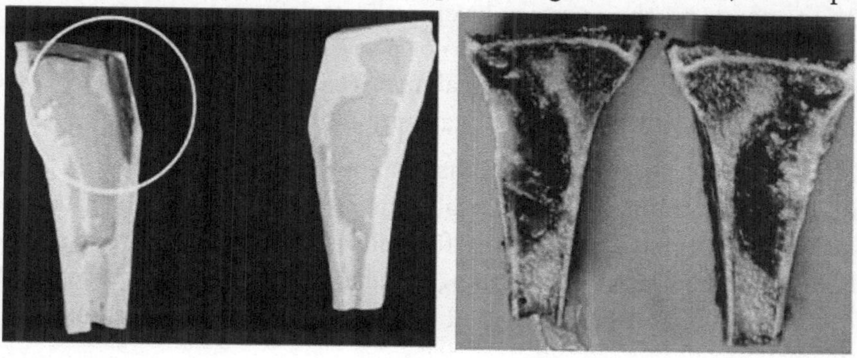

(a) (b)

Figure 7. (a) Osteotomized RP model. (b) Pathological anatomy of tibial osteosarcoma. The similarity between the pathological anatomy and our RP model in colors can be noted, as well as the accuracy of the cut coinciding with the blue osteotomy plane, which corresponds to the safe margin.

4. DISCUSSION

Although the combination of preoperative planning with image-based navigation has already been used in other areas such as maxillofacial surgery [16], spine surgery [17] and cardiac surgery [18], among others, no work could be found in which RP models were cut with saw as part of the training for surgical navigation. As a matter of fact, one technical question that was present before doing this work was whether the RP model could withstand the oscillation of an orthopedic saw without being destroyed. In this experience of only two cases, we probed that the behavior of the material maintained its structure until the end of the experiment.

By taking advantage of the retrospective RP model created, one important challenge was improving the amount of healthy bone preserved compared to that of conventionally performed resections. This means reducing the healthy bone in the resected specimen and leaving a greater volume of healthy bone in the patient, while always maintaining a safe oncological margin. This objective was fulfilled thanks to the 3D control in 3D planning and the 3D control provided by the navigator during the execution of the cut, as it can be observed in the RP model with colors.

Surgical precision, understood as the correspondence between a target osteotomy and an executed osteotomy, is not the only factor to consider when evaluating navigation-assisted surgery. Some complications associated with intraoperative navigation, such as increased procedure time or uncompleted navigation due to technical problems, have shown to decrease as the surgeon team familiarizes with the technology (increasing their total amount of surgeries performed under navigation) [19]. On the contrary, the accuracy level in the registration process appears to be independent of the learning curve and not decreasing with user experience. Local tumor recurrence and non-oncological complications have also been used as parameters to evaluate the benefits of navigation-assisted surgery [20].

The experiment designed for this chapter shows how 3D printing can be applied to build experimental models, specifically for orthopedical oncology, that can be used for multiple applications. In the first place, these models are useful to test and characterize new surgical technologies such as image-based navigation. In the second place, these models can be used by the surgeons for training on particular procedures.

The potential of 3D printed models as surgical training tools for patient-specific procedures has been evidenced for ENT surgery. In a case report where transtemporal tumor drainage assisted with intraoperative navigation was determined as treatment, the preoperatively planned trajectory to access the tumor was executed under navigation first on a 3D model of the patient's skull and then on the real patient. The mean distance between target trajectory and executed trajectory measured in the 3D model was reduced by 73.66% when measured in the real patient. This probes how 3D printed models are a promising method to increase accuracy in surgeries assisted with navigation [21].

3D-printed prototypes are currently gaining accessibility, as 3D printers become more massive and more economical. This manufacturing method is improving both its technical characteristics (such as the speed of printing, resolution and the variety of materials available) and its cost. Therefore, it is reasonable to believe this method can be easily available in clinical practice, turning it into a promising option for the testing of new surgical technologies and procedures.

This work was the first validation experiment of the workflow proposed, which consists of preoperative planning and image-based navigation for orthopedic precision surgery. Subsequent validation and protocolization works have followed this experiment, finally building a routine that is implemented on a weekly basis at the department of computer-assisted surgery at our hospital (Unidad de Cirugía Asistida por Computadora, Hospital Italiano de Buenos Aires) [22].

5. CONCLUSION

The ability to combine different emerging technologies gives important solutions in the field of surgery planning and preoperative surgical design. The 3D prototype as a way of reproducing a surgery is a training model for surgeons interested in knowing the behavior of 3D planning and its reproduction through the navigation of osteotomies in orthopedic oncology. It is possible, in this way, to test surgeries from RP models manufactured from real cases. This method stands out as it is easy to implement and to understand, as well as technically simple and economical.

Likewise, other benefits of this conglomerate of technologies include multiplanar or difficult osteotomies in limb deformities, pre-cast in osteosynthesis plates using the RP models as a guide mold, controlling oncological margins to avoid errors of cut in real surgery and, above all, save surgical time with its cascade of beneficial effects for the surgeon and the patient.

In our hospital department, we were able to adopt computer-assisted surgery for oncologic orthopedics as a standard routine. This includes weekly meetings of specialized medical professionals who perform the preoperative planning for challenging surgical cases, which are then executed intraoperatively under navigation guidance. More than 250 patients have been treated to date following this working protocol, since its introduction 8 years ago. This is an example of how new technologies developed in silico can rapidly reach health care activities.

REFERENCES

1. van de Waterbeemd H, Gifford E. ADMET in silico modelling: Towards prediction paradise? Nature Reviews. Drug Discovery. 2003;2(3):192-204

2. Cho HS et al. Joint-preserving limb salvage surgery under navigation guidance. Journal of Surgical Oncology. 2009;100(3):227-232

3. DiGioia AM 3rd, Jaramaz B, Colgan BD. Computer assisted orthopaedic surgery. Image guided and robotic assistive technologies. Clinical Orthopaedics and Related Research. 1998;354:8-16

4. Rajasekaran S et al. Computer navigation-guided excision of cervical osteoblastoma. European Spine Journal. 2010;19(6):1046-1047

5. Simon DA, Lavallee S. Medical imaging and registration in computer assisted surgery. Clinical Orthopaedics and Related Research. 1998;(354):17-27

6. So TY, Lam YL, Mak KL. Computer-assisted navigation in bone tumor surgery: Seamless workflow model and evolution of technique. Clinical Orthopaedics and Related Research. 2010;468(11):2985-2991

7. Wong KC et al. Precision tumour resection and reconstruction using image-guided computer navigation. Journal of Bone and Joint Surgery. British Volume (London). 2007;89(7):943-947

8. Wong KC, Kumta SM, Antonio GE, Tse LF. Image fusion for computer-assisted bone tumor. Clinical Orthopaedics and Related Research. 2008;466(10):2533-2541

9. Wong K, Kumta S, Tse L, Ng E, Lee K. Image fusion for computer-assisted tumor surgery (CATS). Image Fusion. London, UK: InTechOpen; 2011. pp. 373-390

10. Cho HS, Kang HG, Kim HS, Han I. Computer-assisted sacral tumor resection. A case report. The Journal of Bone and Joint Surgery. American Volume. 2008;90(7):1561-1566

11. Krettek CJG. Computer aided tumor resection in the pelvis. Injury. 2004;35(suppl 1):S-A79-S-A83

12. Gouin F, Paul L, Odri GA, Cartiaux O. Computer-assisted planning and patient-specific instruments for bone tumor resection within the pelvis: A series of 11 patients. Sarcoma. 2014

13. Ritacco LE, Milano FE, Farfalli GL, Ayerza MA. Accuracy of 3-D planning and navigation in bone tumor resection. Orthopedics. 2013;36(7):942-950

14. Muscolo DL et al. Proximal tibia osteoarticular allografts in tumor limb salvage surgery. Clinical Orthopaedics and Related Research. 2010;468(5):1396-1404

15. Roos PJ et al. A new tibial coordinate system improves the precision of anterior-posterior knee laxity measurements: A cadaveric study using roentgen stereophotogrammetric analysis. Journal of Orthopaedic Research. 2005;23(2):327-333

16. Robiony M et al. Accuracy of virtual reality and stereolithographic models in maxillo-facial surgical planning. The Journal of Craniofacial Surgery. 2008;19(2):482-489

17. Lu S et al. A novel computer-assisted drill guide template for lumbar pedicle screw placement: A cadaveric and clinical study. The International Journal of Medical Robotics. 2009;5(2):184-191

18. Sodian R et al. Pediatric cardiac transplantation: Three-dimensional printing of anatomic models for surgical planning of heart transplantation in patients with univentricular heart. The Journal of Thoracic and Cardiovascular Surgery. 2008;136(4):1098-1099

19. Farfalli GL, Albergo JI, Ritacco LE, Ayerza MA, Milano FE, Aponte-Tinao LA. What is the expected learning curve in computer-assisted navigation for bone tumor resection? Clinical Orthopaedics and Related Research. 2017;475(3):668-675

20. Farfalli GL, Albergo JI, Piuzzi NS, Ayerza MA, Muscolo DL, Ritacco LE, Aponte-Tinao LA. Is navigation-guided en bloc resection advantageous compared with Intralesional curettage for locally aggressive bone Tumors? Clinical Orthopaedics and Related Research. 2018;476(3):511-517

21. Ritacco Lucas E, Di Lella Federico, Mancino Axel, Gonzalez Bernaldo de Quiros Fernando, Boccio Carlos, Milano Federico E. 3D printed models and navigation for Skull Base surgery: Case report and virtual validation. Studies in Health Technology and Informatics. 2015;216:1025

22. Ritacco LE, Milano FE, Chao E. Computer-Assisted Musculoskeletal Surgery: Thinking and Executing in 3D. Switzerland: Springer International Publishing; 2016

INDEX